Sound Design for the Visual Storyteller

Sound Design for the Visual Storyteller is an overview of the sound design process for the beginner filmmaker or storyteller, providing the foundational knowledge needed to succeed at utilizing and designing sound for visual stories, films, and even podcasts.

With a focus on television and film, alongside references to podcast, theatre, event, and game sound design, this all-in-one overview begins with the fundamentals of sound and the structure of a professional sound design team, before exploring the practical topics of post-production, creative workflows, and distribution. Supported by a plethora of audio and video examples to demonstrate key concepts, this book guides aspiring sound designers on the power and production value of the well-conceived soundtrack and showcases some of the most effective techniques for getting there.

This is an ideal introduction for storytellers working in a range of contexts, including filmmakers, sound designers, and sound editors, as well as students of sound for film and broadcast.

Christopher D. Anderson is a New England–based sound designer, mixer, sound editor, sound supervisor, and educator. Mr. Anderson has participated in many award-winning programs, working both on set and in post-production. He has been interested in all manner of recorded sound since acquiring a cassette deck at an impressionable age.

Sound Design for the Visual Storyteller

Creating Sound for Visual Media

Christopher D. Anderson

Routledge
Taylor & Francis Group
NEW YORK AND LONDON

Designed cover image: William B. Anderson

First published 2025
by Routledge
605 Third Avenue, New York, NY 10158

and by Routledge
4 Park Square, Milton Park, Abingdon, Oxon, OX14 4RN

Routledge is an imprint of the Taylor & Francis Group, an informa business

Library of Congress Cataloging-in-Publication Data
Names: Anderson, Christopher D., author.
Title: Sound design for the visual storyteller : creating sound for visual media / Christopher D. Anderson.
Description: New York, NY : Routledge, 2025. | Includes index.
Identifiers: LCCN 2024050426 (print) | LCCN 2024050427 (ebook) |
ISBN 9781032446219 (hardback) | ISBN 9781032446202 (paperback) |
ISBN 9781003373094 (ebook)
Subjects: LCSH: Sound design. |
Film soundtracks–Production and direction. |
Sound recordings–Production and direction.
Classification: LCC MT64.S68 A64 2025 (print) | LCC MT64.S68 (ebook) |
DDC 781.5/4–dc23/eng/20241113
LC record available at https://lccn.loc.gov/2024050426
LC ebook record available at https://lccn.loc.gov/2024050427

ISBN: 978-1-032-44621-9 (hbk)
ISBN: 978-1-032-44620-2 (pbk)
ISBN: 978-1-003-37309-4 (ebk)

DOI: 10.4324/9781003373094

Typeset in Helvetica Neue and Optima
by Newgen Publishing UK

Access the Support Material: www.youtube.com/@SDVS.123

To my loving wife and life partner, Julianna. You keep me grounded, but you've also shown me how to fly.

To my children, Ruby, Kevin, Keith, Destiny, and William – you lift me up and keep me humble.

To Janice and John Anderson, Mom and Dad, for your endless support and faith in me to do the right thing and be my own person out here in the world.

Contents

About the Author x
Preface xvii
Acknowledgements xix

Introduction 1

0.1. So What Is Sound Design? 5
0.2. How to Use This Guide 8

1. Foundations and Fundamentals 9

1.1. Sound Wave – Quantifying the Ephemeral 9
1.1.1. Frequency 10
1.1.2. Amplitude 13
1.1.3. Frequency Response 17
1.1.3.1. Masking 22
1.1.4. Velocity 24
1.1.5. Wavelength 30
1.2. Acoustics 31
1.2.1. The Mechanics of Sound – Phase 31
1.2.2. The Mechanics of Sound – Reflections 34
1.2.3. The Reality and Psychology of Sound 38
1.2.3.1. Perception 38
1.2.3.2. HHR – the Realities of Binaural and Selective Hearing 39
1.2.3.3. HHR – Strengths and Weaknesses 41
1.3. Gear – an Overview 42

1.3.1. Microphones and Transducer Types 42
1.3.2. Production Recorders 48
1.3.3. Digital Audio Workstations 54
1.3.4. Signal Processors and Plug-ins 56
1.4. Distribution Systems and Media 62

2. People and Processes 68

2.1. Production Audio – Personnel and How the Good Work Gets Done 68
2.2. Personnel 71
2.3. Equipment 74
2.4. Best Practices in the Field 85
2.5. Post-Production Personnel and How the Work Gets Done 88
2.5.1. Sound Supervisor, Sound Designer 90
2.5.2. Dialogue Editing 92
2.5.3. Foley 94
2.5.4. SFX Editors 97
2.5.5. Predubbing 99
2.5.6. Mixing (or Re-recording or Dubbing) 103
2.6. Tools of the Trade 104
2.6.1. Workstations 104
2.6.2. The Foley Stage 109
2.6.3. The VO Booth and ADR Stage 109
2.6.4. Dubbing Stages 112

3. Creation – How We Do It: Applied Sound Design and Workflows 115

3.1. How Sound Works – Sound Types 115
3.1.1. Diegetic and Non-diegetic Sound 116
3.1.2. Categories of Sound 117
3.1.2.1. Contextual Sound 117
3.1.2.2. Descriptive Sound 119
3.1.2.3. Commentative Sound 120
3.2. Mapping a Plan 122
3.2.1. Various Workflows 123
3.2.1.1. Dialogue 126
3.2.1.2. Sound Effects 129
3.2.1.3. Music 133
3.2.2. Following Through – Transitioning from Production to Post 135
3.2.3. Working in Sound Editorial 137

3.2.4. Coordinating with the Music Department 151
3.3. Envisioning Sound 153
3.3.1. How It All Comes Together 153
3.3.2. Delegating Tasks 154
3.3.3. The Mix Stage 155

4. Distribution – Getting It Out There 161

4.1. Mixing and Mastering for the Medium 161
4.1.1. On Mixing 163
4.1.2. Evaluation and Specifications 166
4.1.2.1. The CALM Act, Level Specs, and How We Got to This Place 166
4.2. Formatting and Deliverables 171
4.2.1. Distribution Formats 174
4.2.2. Broadcast, Streaming 176
4.2.3. Film 177
4.2.4. Podcast and Streaming Platforms 178
4.2.5. VR and Gaming 179
4.3. Summing It up (Pun Intended) 179

5. That's a Wrap – an Afterword 180

6. A Semi-Complete Glossary of Terms 183

6.1. (Some) Terms You'll Want to Know 183
6.2. References and Other Source Material 202
6.2.1. Educational and Informational 202
6.2.2. Sound Guilds and Unions 202
6.2.3. Sound Effects Libraries 203

Index 205

About the Author

I didn't expect to be in this place.

As a young boy growing up in a very rural setting, I think I was a bit of a bafflement to my parents – there was me, a creative, somewhat musically inclined kid, born to parents who were steeped in careers in healthcare and health sciences. I thought any medical stuff was icky, preferring to expend my attentions and energies on mechanics and electronics. I was that kid who was always dismantling things, from bicycles to radios to the lawnmower to minibikes to my mom's car, much to her chagrin. I would then advance to electronic circuits, electric guitars, basses, and amplifiers – tinkering, modifying, and essentially hot-rodding them to get them to perform better.

This was all in the name of my intense need to know how things work. Usually, these things could be successfully reassembled, sometimes with only varying degrees of success – at least getting them back together and working as they had been. Bonus points if I could do it without any extra parts lying around.

I began playing guitar at 6 or 7 years old, when I found my dad's old acoustic guitar in a closet. The action was abysmal, the strings were way too high off the fretboard to be comfortable for my young hands, but I was persistent and serious enough about it that my mom signed me up for lessons. I circled away from the guitar and back again over the years, often more interested in the technology around it and the crazy sounds it could make than actually mastering or playing the instrument itself. More disassembly followed, and more circuits were created … or destroyed.

A very early clue to where my life was headed could be found in my fascination with producing home recordings. My dad was a doctor and received audiocassettes marketing the latest innovations of pharmaceutical companies in an effort to get him (and all doctors) to prescribe their newest product lines. I guess the fact that it was all on cassette puts a date on my experiences. Blank audiocassettes were an expensive commodity for a 10-year-old with no income, but once I figured out how to defeat the record protection tabs on those free cassettes, they became a near-endless media source. I would use and reuse them for my own (sometimes) nefarious ends. My dad and brothers and I would produce satirical "commercials", mostly touting the benefits of Mom's cooking, but also promoting fictitious products and services of our own invention. We also created some rather epic sound stories, developed while I was discovering new modes of storytelling in various comedy show recordings.

Old Jack Benny radio shows were a favorite; I also had the eye-opening experience of listening to a Lenny Bruce album at a friend's house, as well as early George Carlin. I was fortunate enough to pick up a few Firesign Theater LPs at a neighbor's yard sale, and my dad was a huge Monty Python fan – we had all the Python LPs, and would also watch the television show regularly. I would also tune in weekly to the *CBS Radio Mystery Theater*, hosted by the venerable E. G. Marshall – that show's almost cinematic approach to sound design was a source of endless fascination to me, and I could actually see the story in my mind's eye.

I was also intrigued by the idea of radio personalities – specifically, Rick Dees, Casey Kasem, and Wolfman Jack. With imitation being the sincerest form of flattery, I would produce my own radio show with whatever Top 40 music I could grab off the airwaves or cue out of my parents' collection of LPs. This was entirely for my sense of entertainment, those tapes never saw the light of day, and consequently, my career as a radio personality was never fully realized – but in hindsight, it's not hard to see this burgeoning career of mine coming on.

There was one epic failed film project in middle school that set the scene for the career that would follow when some friends and I got a hold of an 8mm film camera. There was no way to record sound on those, so we used my trusty cassette recorder and those recycled cassettes for the audio, inadvertently reinventing the double-system recording setup – but then quickly realizing there was no practical way for us to re-sync those takes back together, particularly with the absence of any kind of system or process to help align them later. And we also kind of forgot (or more likely, just didn't know) anything about the editing process that should have followed. We had fun shooting our scenes, though, and our loosely scripted, linearly shot narrative made sense to us, at least, even if we couldn't adequately assemble it when we were done. Our teacher was very kind and gave us a good grade anyway, based only on seeing some of our raw footage and after enduring some very disjointed in-class playback efforts – an A for effort, I suppose – but that first failed attempt at a dailies session was the germ of an idea for me that ultimately led to my future in audio post-production.

Flashing forward a few years from there, I played bass in a couple of different high school garage bands. We were only ever OK – I'll admit the bass-playing there was merely adequate, at best, but I was fortunate enough to be in with a truly talented group of musicians, most of whom still play at least semi-professionally. I made up in volume for what I didn't have in talent. We were mainly playing cover songs at parties, covering the frat house circuit at the local college, some high school dances – and somehow it always ended up that I'd be the one who managed the PA system. It may have seemed like drawing the short straw for anyone else, but it wasn't a chore to me – I loved it and wouldn't have had it any other way. It was hard to articulate at the time, but when I was running that PA board, it felt like the whole band was my instrument.

I also really liked the concept of a recording studio. Never having been in one before, it was an abstraction at best – only slightly supported by what I could catch in the media and the few trade magazines that made their way to the relative hinterlands of my rural upbringing. Just before my senior year, my band and I lucked out and got into a nice, professional recording studio near Boston, an experience that quite literally changed my life. I was now on a mission and my life plan had been revealed to me. From there, it was off to the Berklee College of Music, where I joined the inaugural class of Berklee's brand-new Music Production and Engineering program under the direction of Wayne Wadhams and the legendary producer Don Puluse. By all practical measures, I did well in that program, and my career path was obvious – to me, anyway.

My expectation had always been to head west and do big west coast music sessions, but life had other plans. Ultimately, I only got as far west as Pittsburgh, where I became (improbably) the chief audio engineer in a brand-new video post-production house, Production Masters, Inc. I was pretty sure I had won the lottery here, as even the most aspiring rookie engineers straight out of Berklee weren't likely to land this kind of gig as their first (or even second) job. It also came with a heaping dose of imposter syndrome, and I set out to prove myself in my new position.

I had shown an aptitude for engineering and synchronization systems, so one of my Berklee mentors, Robin Coxe, taught me how to run our department's then-state-of-the-art synchronization system. This course of instruction was pretty much off the books, as there was no formal course in post-production sound or sound synchronization at the time – but at the same time, someone in the Music Production and Engineering Department needed to know how to use it to support the film scoring sessions that came through our studios. Berklee also has an excellent Film Scoring department, but there were very few of us who knew the systems well enough to run those sessions.

Our typical setup was usually a multitrack analog tape work deck, with reference video on timecoded 3/4" U-Matic tape, with effects and music flown in from a third 1/4" deck, CD, or (it being a music college with its own Film Scoring department) actual live musicians playing to a synchronized click track. The synchronizing system was slow, it was very touchy, and you had to develop a sort of sixth sense to know what it might be about to do to you and your tapes – but there was a sweet spot where it worked well. The skills that Production Masters was looking for were in short supply in those early days.

Flash forward a couple years, and I got an offer I couldn't refuse and came back to Boston, where I would ultimately stay. I've worked with many different analog and digital systems along the way. Seeing these systems grow and evolve was gratifying – particularly when some of my input was considered in their development. The technology was growing and improving quickly, and many of us could see the potential even early on – the future was bright, and this game-changing technology was improving with each software iteration.

Through all of this, the fact remains that, no matter the technology in play, the art of it always was – and will continue to be – the key. While the actual means to arriving at a wonderfully realized soundtrack might be different and evolving over time and in a state of constant growth and change, the actual theory, fundamentals, and practices that get us to that place are still essentially unchanged, even after all these years. The result is – ideally – an artfully executed soundtrack that supports the film and furthers the telling of the story, bringing it to another, higher level.

Since my inauspicious beginnings, I've worked in more than a few post houses, co-owned a recording studio, and I currently run my own company as well. Things were going great, but an unexpected twist in this career trajectory came when Berklee, my alma mater, asked me back to teach an audio-for-video engineering course. I hadn't trained as a teacher, it hadn't been on my radar at all – in fact, I wasn't even that far removed from the college student experience myself at that point. I'd had some incredible teachers and mentors along the way, some credited in this book – but in the face of these excellent examples in teaching and mentorship, I wasn't sure I could ever rise to their level. As it turned out, my Berklee teaching experience was terrific, and ultimately a warm-up for bigger and better things to come. Not long after, a couple of good friends asked me if I'd consider teaching a sound course at Boston University's College of Communication in the Department of Film and Television. I was honored, but again unsure of what I could bring to the college experience.

It turns out, there is a desire for those hard-fought, hard-won skills I've developed over the years. But while I do have a base of knowledge and experience to impart, I find that I learn at least as much from my students as I have given to them – maybe more. I know that's a hard concept to grasp when this equation is being seen from the student's point of view, where you're working hard to earn a good grade while you learn new and likely unfamiliar concepts, under deadline – but this process of teaching requires me to have my knowledge organized, together, and accessible. While I do need to know the material I'm talking about, it's not worth much to my students if I'm not able to package it in a way that is understandable to them.

Even more importantly, I need to continue learning for myself. Teaching has taught me to be more organized in my approach to sound work, it has forced me to articulate what it is that I do in my day-to-day life as a sound designer, and then to present, in a cogent way, what works and (so much more importantly) why. That process of organizing thoughts and processes into something relatable also allows me to re-examine the way I do things, and then – as is my nature – look for ways to streamline, refine, and maximize whatever it is.

So much of what we do in sound feels subjective and seems to come about by feel, intuition, and even serendipity, but there is also a lot of necessary science behind it. Knowing what works and why further helps seed future serendipitous happenings – a kind of "applied serendipity", which is just opening the door to more happy accidents. Those happy accidents can, do, and will happen when you facilitate them properly. But for me, in the end, that process of questioning the whys, hows, and wherefores of all that I do is only the first step. Assimilating it and figuring out how to relate that knowledge and skillset in a way that transfers that knowledge remains the biggest challenge.

I hope that you will use this text as a treatise on soundtrack design and implementation, but also as a reference that you can revisit again and again. One should never feel they're too experienced, too old, or too jaded to have room for more learning in their life, and my aspiration here is to help lay out and articulate a path for you into our sound world.

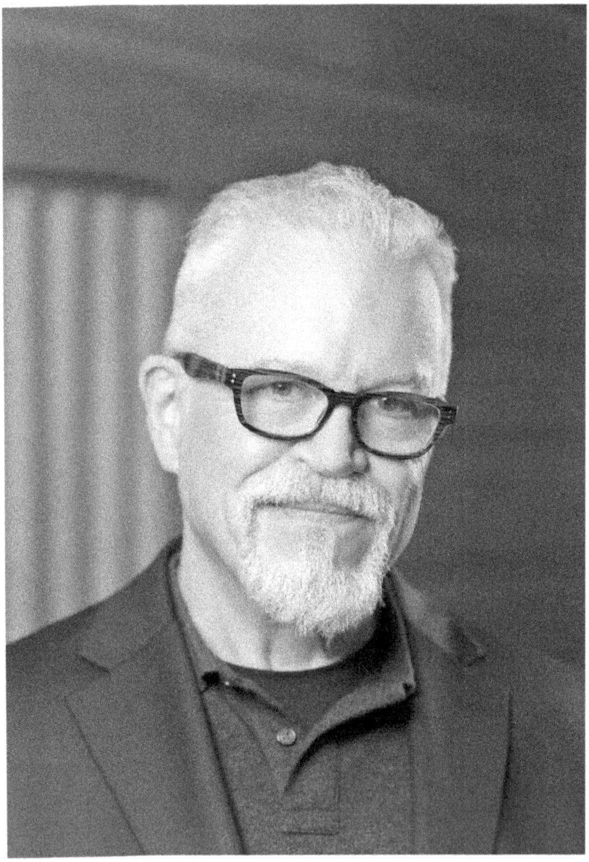

Mike Spencer Photography

Preface

I'm an unlikely candidate to be writing a textbook.

I have taught at various institutions, most notably Boston University's (BU's) College of Communication, where I currently teach a sound course in the Department of Film and Television. BU is running a more traditional film school here, first and foremost, and the sound curriculum is therefore not as complete as some other programs might have it, as it's simply not the focus or the primary mission. We're raising holistic filmmakers and storytellers, as I see it, and they need to be versed in all aspects of filmmaking when we're done with them. Within that, I have a unique mission – to give my students as much exposure as possible to the world of sound and its various practices as they relate to film. All of it designed to fit within the context of a one-semester course that meets once a week.

So, the simplest rationale for me writing this text, quite simply, is that I wanted a book that would be better suited to my students, tailor-made for the beginning filmmaker and aspiring visual storyteller that also embodied our do-it-yourself ethic in the Film and Television Department at BU. I wanted to give them the tools to succeed at sound, while also – hopefully – demystifying some of it for them.

While there are a few texts that I could have forced into working for my course, nearly all of them are almost *too* comprehensive for my classes. I'm constantly working on how to provide the best sound overview, while also providing adequate background and grounding in facts, concepts, and practices to a group that has its eye on bigger, broader prizes. It was upon identifying this need that I decided to undertake the task of writing a book about it, assembling several years' worth of experience and curating materials gleaned over the years.

It should be noted that I've been talking about writing this book for at least 20 years now, so it's been a very long incubation period.

In terms of how to present the material, well ... I am of the mind that filmmaking is a team sport, and everyone on that team should be working in service of the story – a simple fact that sadly falls by the wayside sometimes. I've gotten caught up in the process myself, overwhelmed by the technology and the choices it affords, and it can become difficult to see the forest for the trees.

My course – and by extension, this text – are both designed to help point the aspiring sound designer toward the mission of collaborative, creative sound.

Every sound provides us with an opportunity to further the story, from the incidentals – footsteps, the sounds certain props make, and so forth – all the way up to the more elaborate and designed sounds, like monsters and dinosaurs and aliens and spaceships. It is in our deliberate, thoughtful choices that the story achieves its best, most fully realized version of itself.

Researching and pulling materials together also provided a great context to talk with many of my peers about their art and their pursuit of sound excellence – an excuse to "talk shop" while on the clock, essentially, which isn't always something we like to do when we get together socially. But under the pretense of research and pulling information together for this text, it provided some great insights along the way. One interesting aspect was learning how my peers got into the field – as outlined earlier (About the Author), one can see signs of this career of mine emerging in my life from a relatively early age. No two of us followed the same path, however, and different aptitudes and circumstances come together in different ways to lead us to whatever place we're in. I hope that some of those stories might help inspire you to find your direction, and maybe help you maximize certain tendencies in yourself along the way.

I knew writing a book wouldn't be like preparing for a class lecture, but I somehow thought my class lecture notes, slides, and other materials would translate better into book form. In the journey of self-discovery that is now this book, I discovered that those notes, while helpful, were little more than an expanded list of talking points that needed me to expound upon them for any of the material to make any sense at all. This is certainly something I do when in class, it's how three pages of bullet points always manage to take up the better part of three hours; it's the information contained within those points that takes up all the time. Each line item turns into a paragraph – or three, or five! – of text before it can even begin to make sense, and only then after relentless editing and seeking input and more editing, re-reading, and reorganizing. I didn't expect any of this process to be easy, and it certainly wasn't – but I'm not sure I expected it to be quite so intensive, either.

In the end, I'm glad I did it – it's a labor of love and art, but it's also my guide to hacks and shortcuts I've learned over the years, and ultimately, ideally, how best to get things done. As happens in my class, every term, I find I have learned at least as much about myself as what got put down on the pages of the book, and for that, I'm grateful.

Acknowledgements

Again, none of this would be possible without the love, understanding, and support of family.

Thanks and a lot of credit go to my Mom and Dad – you didn't always understand what I was about, but you supported me anyway, no questions asked. A few questions, maybe, and largely just the important ones. Like, "How much is this going to cost?", and "Can you really make a living doing *that*?"

On the professional side, there are many (so many!) whose generosity and willingness to share helped me get to where I am today. I'd start with Brad Terry and Steve Grover, two amazing jazz musicians who taught me how to really, truly listen, way before I knew I wanted to be any kind of audio professional. There were many late-night jazz listening parties, sneaking me and my friends into their gigs through the back door of whatever club they were playing, as long as we were quiet and didn't implicate them in any of our shenanigans. Mainly they taught me a key life skill through their mentorship – how to improvise, both in music and in life. All of this served to deepen my appreciation for the style, but also, and most importantly, they showed me how to hear inside the music to understand what was going on with the arrangement, the composition, and even the production of the piece.

Joanna Jones, my high school music and chorus director, a brilliant leader and teacher – I learned so much from her about the mechanics of good speaking and singing, but also about what true leadership and musical mastery look like.

My Berklee College of Music career would have gone nowhere were it not for the strength and encouragement I received from Robin Coxe. She taught me everything I could absorb about the fundamentals of audio engineering, but also audio-for-video synchronization systems when that technology was still very new to us in the analog tape world. This, as well as imparting a strong appreciation for blue-colored cocktails, gave me an entire suite of knowledge and appreciation, for all of which I am forever grateful.

Sue Hartford, who took Robin's good advice and hired me right out of Berklee. Sue taught me the business and administration side of things, as well as the invaluable experience and practice of working for and with clients in the real world. She led by great example and provided me with a masterclass in directing singers and voice talent – all while helping me apply my newfound education and theories and giving me the room I needed to grow into my new profession. It was an exciting time and a great first job, and I also learned all this while we navigated the earliest beginnings of digital audio technology together.

Jim Anderson (no relation), who showed me how to be a decent human in the often hard and fickle world of this strange audio business we find ourselves in, even in the face of intolerable demands, demanding clients, crazy rockstars, unappreciative and meddling studio owners, and some of the most unique personalities I had ever come across. Coping with and managing this craft start with a good laugh now and then, and his insights and empathy were essential and invaluable.

Mary Jane Doherty, Sam Kauffmann, Charles Merzbacher, and Paul Schneider in the Film and Television Department at the College of Communication at Boston University deserve a shout-out for all they've done in supporting me over the years. Mary Jane, being a fierce sound advocate, talked me into teaching my course many years ago. Sam, Paul, and Charles have all been more than kind and have kept me encouraged and supported.

Tim Mangini, Technical Director at WGBH in Boston, took a chance and showed a lot of faith in me on more than one occasion. A man whose wit and wisdom far exceed my own, who leads by great example, has the ears of an eagle, and who put together an amazing, unflappable team that I'm grateful and honored to be a part of. I'm indebted for the insights and input he's provided in the creation of this text here.

I'd also like to thank my unions and guild – IATSE 700, the Motion Picture Editors Guild, IATSE Local 481, New England Stage Mechanics, and NABET Local 51018 – for their support and fighting the good fight for all of us doing the work every day. I thank the Cinema Audio Society (CAS) for their recognition and for allowing me to be a part of their ranks, it's been an honor, and I wear that CAS at the end of my name with great pride.

To the many eyes, ears, inputs, and proofreads that helped get this thing off the ground in the first place – Tom Williams, CAS, Jim Sullivan, Mark Mangini, Tony Crowe, and Mark McBryde. Tara Blume and Mark McBryde for their help with the Foley section, Ryan Maguire for his insights on the behind-the-scenes work of a Foley stage, Ed Grenga for his thoughts and insights on music in film, and so many others for their thoughts and views from both sides of the camera.

On the technical and administrative front, my editors, Evie Evans, Emily Tagg, and Hannah Rowe, deserve a lot of kudos for their patience with my neophyte ways. Drew Stanley, my copy editor, for polishing my writings and making them accessible and articulate for all. David Eber helped me through the technical inner workings of the publishing world and gave me some great insights into the business aspects of what makes a book like this happen.

Many, many thanks to dear friend and publishing insider person Christina Smith DeYoung for sharing her thoughts and expertise around the mysterious world of publishing, and then coaching me along and helping me find the courage and permission to just write, dammit!

Finally, and most importantly of all, my wife and partner in life, Julianna. Your unquestioning support, endless patience, and enduring encouragement were the push I needed and the key to making this work happen at all.

To my children, Ruby, Kevin, Keith, Destiny, and William – you all are an honor and a blessing. I finally get what they mean when they say parenting is a gift – I am certain I've learned at least as much from you as I've given you, and I am beyond grateful and proud to be your Dad.

Introduction

In the early days of my career, still feeling my way and doing my best to earn a (good) reputation in the industry, I took on some adjunct teaching jobs. It would be fun, and I could share insights into my recording studio day job as an audio post-production specialist. It was equal parts curiosity, a need to be challenged, and the idea of giving back; the hope that I could provide some guidance for aspiring filmmakers on their way to a successful career. I'm not sure what I expected, but it didn't go exactly the way I thought it would.

Over the years, it has proven to be challenging, gratifying, humbling, and inspiring. Then, inspiration struck, in the form of writing my very own, tailor-made textbook. I had many texts to choose from, but none seemed quite right for the body of students I was working with. These were not future sound engineers; they were storytellers first and foremost, and their main interest in sound was as a pathway to a better telling of their stories.

One great humbling moment came to me in a story one of those students told me about her dad and his responses to some of the courses she was taking – specifically mine. As he looked over her schedule for the term, he came to my Sound Design for Film course, and snorted, "Sound design? Was basket weaving full?"

No disrespect meant to aspiring basket weavers, he was only indicating the same disconnect that many people feel, and to him, basket weaving seemed to have as much to do with her film major as this so-called sound design class.

That dad's confusion does point out how this idea of *Sound Design* almost feels oxymoronic; we are familiar with design as a visual term, but how would one design sound waves? One might as well try to impose a design on the air we breathe – the term feels esoteric at best.

DOI: 10.4324/9781003373094-1

Early in my career, it seemed every sound engineer I met had granted themselves the title of Sound Designer. I was initially a little awestruck, wondering what key bit of wisdom these wise practitioners of sound had that I had yet to achieve that would allow them to take such a presumptuous title.

So, what, exactly, is sound design?

First, it's necessary to manage expectations and establish a few common denominators, and perhaps back up for a minute to consider everything that goes into the term. It begins by first acknowledging that the vessel for any sound design is the soundtrack itself. From there, we find that the definition of a soundtrack can vary widely and can often even depend on the context of the individual considering the term. For the most part, the film-viewing, streamer binge-watching public would probably tell you that it's the music score that accompanies a film, particularly if popular music is part of that score.

For those of us doing the work of visual storytelling – meaning filmmaking, television production, online content producers, and visual media in all its manifestations – the term "soundtrack" is a broader, all-encompassing term that includes anything and everything to do with sound. It covers all aspects of sound being applied to the visual work, including – but not at all limited to – the music. It includes the dialogue track, the sound effects choices and placement (the *design* part of the term *sound design*), and yes – also the music. These sound elements, combined in a sort of recipe of sound, are put into service together to help tell the visual story. How we put that soundtrack together – in a narrative-supporting, story-furthering, and aesthetically pleasing manner – is why you're here now.

Of all the crafts and trades based in and around visual storytelling, sound is one of the hardest to quantify – probably because it arrives in so many forms, with the elements of dialogue, sound effects, and music being just the start; they're just the three macro-categories we begin with.

Sound is an integral part of the human listening experience for most of us, going back to our time in the womb. From there, we continue to experience it continuously, and in an immersive state; it's a constant in our lives. Because of this, we tend to take it for granted and overlook its importance and implications for soundtrack design, as well as its usefulness as a storytelling device. It's all around us and never stops; we are constantly experiencing new sound information every waking second of our daily lives. How we perceive – or sometimes, choose to ignore – all those sounds around us can lead to an examination of our sensitivity to sound, leading us to better understand the role it plays in our perception world.

Applying sound with mindful intent and in support of storytelling is the next level of refinement of the definition of sound design. Whether it's a story we're hearing as spoken word or some form of recorded visual storytelling, the sound that is applied in accompaniment serves to support and strengthen the impact of the story being told. All aspects of sound come into play here, whether we're aware of it or not.

Sound perception can be influenced by a performance – for instance, a spoken-word storyteller might choose to manipulate the inflections and projections of their voice to get a point across, varying their performance to fit a particular narrative. Certainly, we can see how accomplished actors have honed this to the highest art in the course of pursuing their craft. After all, an exciting story is much less so if told in a flat monotone, and an intimate conversation loses all intimacy when it's shouted at us.

We can see in history that drums, other musical instruments, and all manner of sound props have been applied to further the storytelling cause. We begin to see how a story might be received through the effective application of certain sounds, and it is here we get our first hints of a definition for that elusive term "sound design".

From this point, we might find that other, more articulate accompanying sounds are required to further tell our story. We can begin with the incidental sounds that illustrate and even showcase the action being portrayed – these can be reality-based, visually driven sounds like footsteps and the handling of items we see on screen, as well as the sounds of the environment and setting where our story takes place. Many stories, from film to television to modern-day podcasts, use these techniques to help create an effective narrative, but their roots are in the pre-electronic days of theater, vaudeville, and then, later, radio plays and television dramas of the last century.

As we have advanced in our storytelling sophistication, our focus is often on attempting to bring the listener or viewer into the contrived world of our story – in short, to help them suspend their disbelief long enough for them to forget they're on the outside, looking in. Good sound design is there to help create an immersive world with sound to an extent that makes the viewer or listener feel like they're also inhabiting the world of the story. To accomplish such a feat is no small thing – it often requires attention to even the smallest detail, as well as a level of technical sophistication. The execution of high production value ideals and concepts will – ideally – help pull it all off.

To be sure, technology is an essential tool in the cause – it is a means of furthering the exposition of our story and making key points, while, at the same time, immersing our listener and allowing our audience to suspend their disbelief long enough to feel like they've joined the world of that story.

The point here is that sound should be in service to the story, from dialogue – perhaps the most obvious element of the sound story – to the sonic environment where the story takes place, to the choices of music we make. Every sound has an opportunity to help tell the story and make an impact on the listener.

At the same time, technology should only be applied in support of creating and furthering the soundtrack, not to get in the way of it.

One recurring theme for sound artists working in support of visual storytelling is that, after the relentless pursuit of furthering the goal and the art behind the story being told, it should also happen in a way that does not call attention to its process. The choices of the sounds and sound design technique should remain as transparent to the viewer/listener as possible. Like the "fourth wall" concept in film, sound design should help create and maintain the always-fragile illusion of the world of the story. That is a spell that is easily broken, by inadequate or subpar sound, inappropriate or misplaced sound effects or music, or just a poorly mixed soundtrack – anything that's just a little bit "off" effectively pulls the viewer out of the moment and breaks that illusion.

It may seem an odd parallel to draw, but bear with me for a moment here: physicians have the Hippocratic Oath, the primary tenet being "first, do no harm". While it's hard to imagine a soundtrack inflicting the kind of harm on anyone a malpracticing physician might (sound design rarely has life-or-death consequences), we sound artists know we have done our best work when the soundtrack doesn't do the "harm" of calling attention to itself in any overt way. We can certainly appreciate the spectacular sound effects that go into a typical Hollywood blockbuster, but in the end, that work happens on a more visceral level; it's felt more than noticed, it's there to provide seamless support to the action on-screen and further the story along, while also conveying all the main points of the narrative – and doing all of that work without calling undue or unwanted attention to itself. In the end, the best and most effective soundtracks are the ones that enhance the story but which no one directly notices.

The purpose of this book is to showcase some best practices and techniques but also to help the reader understand that those techniques can be adapted to fit the scope and the scale of the project, the budget, and even the genre. As we move forward in *Sound Design for the Visual Storyteller*, we will consider what goes into doing good sound work, from the software to the personnel, processes, and facilities where the work takes place, as we develop and devise a tailored approach and workflow. There can be many variables, but in the end, the practice, techniques, and technology are just a part of it, with the storyteller's creativity being the thread that holds it all together.

Individual processes and modes of working can (and should) be a highly personalized and specialized thing. Sound practitioners have preferred styles of working, with all the equipment, procedures, applications, and practices we are most comfortable with and have honed over time, but they begin in a simple premise – achieving the best sound possible and bending that sound to do the work of telling the best version of a story.

In the end, after all these years of filmmaking and the many centuries of storytelling that came before that, the idea of the story itself and relating the most effective experience is still at the heart of the matter. It's the means of delivery that varies.

Technology is not a panacea, but it's important to have some technical expertise under one's belt before getting into any kind of sound practice, so the first chapters focus on the foundational aspects of sound and sound gathering. Being an admitted technophobe, I find that technology works best for me when it stays out of my way. In the end, I just want to make good sound, with my hands on faders, and all the processing and power at my beck and call – the less intuitive the technological process becomes, the worse it works for me. I know I'm not alone in this, so my stance in this text is decidedly less technical than many other works out there.

Chapter 1 introduces you to the physics and idea of sound, covering theory, then basic practices, and moving on to the gear typically used and the workflows to which it is most often applied. It's a natural progression and will instill the foundation you need to get to the fun stuff.

0.1. SO WHAT IS SOUND DESIGN?

It helps to clarify some terms first.

We all have some idea of what sound design is and what it means, while at the same time understanding it doesn't create something as tangible as basket weaving.

Good sound design is like art. We may only know it when we hear it, even if we can't easily quantify or describe it articulately.

That is not entirely helpful as a working definition, so some history to help …

Whereas sound and music have been supporting elements in film since the first talkies, the term "sound design" first entered the storyteller's domain and lexicon in the late 1960s with Dan Dugan's work for the American Conservatory Theater's 1968–1969 season, which saw him employ 3 tape decks and different speaker arrays to "design" the sound part of the theater experience for the audience.

It's important to note that the greatest challenge of theater sound is the unpredictability of it at times – performers will vary their pacing as they hone their performances, but also according to the audience reaction … and sometimes, they just miss their cues. It's likely that, over a series of performances of the same show, it will change and evolve in subtle (and maybe some unsubtle) ways.

The world of theater is very different from film, where sound elements are applied to just one "performance", that being the final cut of the visual story. Visual elements are essentially "locked" into place.

The term *sound design* came into wider use in film after Walter Murch's work in *Apocalypse Now* (Figure 0.1). As history notes, Mr. Murch worked on portions of the film edit, but it is in his sound work that we can discover a big part of what made this film so important for sound.

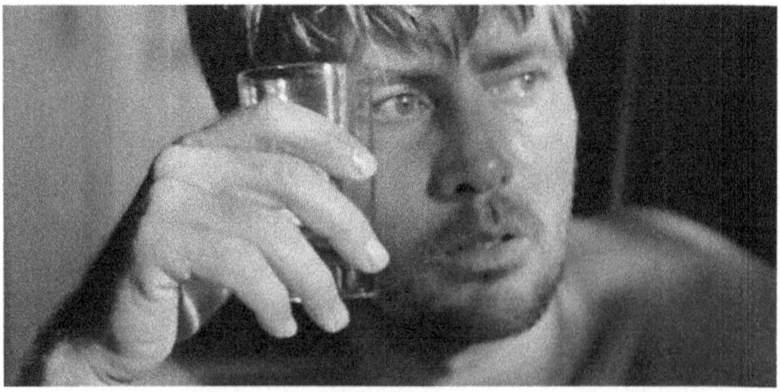

FIGURE 0.1 Screenshot from *Apocalypse Now*

Murch wanted to create an immersive sound field to support the visual action – from the claustrophobic quality of Martin Sheen's narration to the demand for authenticity in the sounds of the Vietnam War. He used actual sounds of Bell UH-1 "Huey" helicopters, showcased the differences between the M16 rifle of the US troops versus the clatter-y sound of the Chinese-made AK 47 rifle favored by the Vietnamese, and so on. He rejected the then-common practice of using "canned" sounds – that is, sounds from pre-recorded sound effects libraries – as he built what he called "sound collages".

Murch's pioneering work on *Apocalypse* and the newly minted title of Sound Designer helped to set the stage for what was to come.

It was a significant and powerful move on Murch's part, as his efforts helped bring sound to the forefront of American cinema, making it less of an afterthought and more of a supporting element in the story – and the American studio system began to take notice. The idea of sound design and the role of sound designer both gained traction in the sound profession, and the studios began to realize that good sound had high production value.

In more recent times, we see less use of the title, and the idea of sound design has become more conceptual. Consequently, we see the title of Sound Supervisor coming back into wider use in film, while at the same time, the title of Sound Designer has expanded into the areas of music composition, game sound, and even sound reinforcement practices.

There are many definitions, but sound design is when all the sound elements work together to help further the narrative while doing it in a coherent, artful, intentioned way. Good, story-supporting sound shouldn't distract from the visual and call attention to itself, but should also be free to shine, in certain circumstances. It can be over the top or subtle, dramatic and striking, or rich and textured, but all in furtherance of the narrative.

As in all things filmmaking, there are exceptions and qualifications and terms and conditions to apply to all that. There are no hard and fast rules here and, as in life, some rules are meant to be broken – or at least bent a little. Some of the biggest successes and advances in film have come about when existing conventions were ignored.

One issue in all this is that the average viewer doesn't often perceive sound design as a production. As mentioned earlier, we sound folk must follow the precept of "first do no harm", and potential harms that can be done to the story being told should not be understated. This happens when the soundtrack calls attention to itself, by way of being inadequate or substandard, or even by being over-designed and too over-the-top for the story at hand. A sound treatment becomes inappropriate when that soundtrack breaks the spell the story is attempting to cast on the viewer. By calling undue attention to itself, that soundtrack can break the viewer's immersion into the story, which effectively pulls them out of the moment and ruins the illusion we were trying to create in the first place.

Neither end of the spectrum is doing our story any good, and so the tenet stands – the real mission of the soundtrack is to provide support for the visual story.

Lastly, a little clarity on the use of the term "storyteller", a more modern term I prefer to use, as opposed to filmmaker. These days, many more visual stories are being told without a frame of film ever being exposed, so the term "filmmaker" might be inadequate to cover all aspects of visual storytelling, which can range from online content to social media stories, but also includes audio stories and podcasts as well. It is highly debatable, at this point, whether the "filmmaker" moniker is at or even near the end of its shelf life. My mission is not to hasten its demise but to simply broaden the application of this practice of best sound to storytelling, in all its modes. Good sound design goes far beyond the film application.

I'm certainly no maven of cultural terms and their use and administration, so what ultimately happens with the title is not mine to decide. My only point is that we have so many more ways of expressing and expositing our stories these days than just film or video projected onto a wall. Even before the pandemic, we were seeing exponential growth in streaming platforms, and the online model of distribution has only gotten stronger because of that time. From social media like TikTok, Instagram, and YouTube, to the streaming revolution started by Netflix, we are still seeing the effects of that explosion of platforms and outlets for programming – and with that comes a nearly insatiable need for content.

It's a wonderful time to be a visual or auditory storyteller, in other words, as there are myriad pathways to getting those stories in front of so many. As I said, good sound is a great production value, so I would pose the question – what better way is there to get your storytelling works noticed?

0.2. HOW TO USE THIS GUIDE

This guide works in two ways – first and most obviously as a conventional, linear chapter book following a natural progression, where A is followed by B, followed by C, and so forth, with the book taking you through the process of sound work on a typical visual story production. But those chapters also stand on their own, to some extent. The intent is that you use this text as an ever-ready reference guide for whatever process is to come.

Also included is a glossary of terms a storyteller may need to know. This is to flesh out some of the terms referenced within the text, rather than being meant as a comprehensive guide. In this day of instant-access knowledge, it's somewhat redundant to reinvent that process, not to mention space-consuming – such a work could be a whole new text on its own. There are certainly better guides and glossaries than this one for such information gathering.

My best hope is that this guide will serve as a gateway to the broader sound world and be the beginning of your sound journey, and that this will occupy a place on your bookshelf – virtual or analog – for many years into your career as a visual storyteller.

Sound is timeless, ephemeral, and sometimes hard to quantify. It requires a medium – in our case, air – to be experienced, and for all our countless ways to grasp and capture and analyze it, it still defies any practical tangibility. Compare this to visual media, where we can freeze a frame of film or video and take a good, long look at it – we can dissect it, analyze it, and manipulate it to specific ends. Sound defies that kind of scrutiny; without the ability to control the time axis and that still-frame capture capability of video, we can't pause sound in the same way we can with the visual. Because we humans experience time linearly, our sound is best experienced in linear time – without the two axes of time and space, our ability to better "look" at our sound falls apart, and we fail to take in the entire sound experience. Good sound cannot happen in a vacuum, it requires a medium to propagate and be experienced and heard, but there is also the vacuum of the unknown. The visual storyteller needs to understand how to gather, store, and manipulate sound, while also intuiting how sound behaves and interacts with other elements of the storytelling exercise.

All with an eye (or ear) to how we humans experience and respond to it.

Easy, right?

CHAPTER ONE

Foundations and Fundamentals

1.1. SOUND WAVE – QUANTIFYING THE EPHEMERAL

The tricky part of measuring sound is that it's all under a sort of disclaimer. Certain terms, contexts, and conditions may apply, meaning that many aspects of sound perception are subject to certain other, less quantifiable properties – some are environmental, some inherent in an individual's hearing response, perception, and interpretation. The truth of it is that every individual's perception of sound will vary at least slightly from the next. Sometimes, these variations can be significant; sometimes, they're only measurable in the smallest of ways.

Each person's perception of sound is unique, with variables ranging from an individual's genetics and biology to their environment, present and past. It can even come down to an individual's personal inclinations, biases, tendencies, and tastes. All these factors come together under the header of Human Hearing Response (or HHR, we sound people like our acronyms), a place of seemingly as many variables as absolutes. On top of all the variables that are a part of HHR, there are also quantifiable variations out here in the physical world as well. These physical phenomena and interactions affect how we perceive and experience sound.

In the end, measuring sound has concurrent factors – at least on the surface, it's a relatively straightforward measurement. It's only physics, after all, and science remains reassuringly quantifiable. But at the same time, it can get complicated, particularly when we attempt to quantify individual hearing response through the filter of simply being human. At this point, the variations accumulate, so to get to a place of accurately portraying and quantifying sound, all factors must be considered. We first rely on all technical and theoretical means available to us, take some leaps of faith, and ultimately examine some of

DOI: 10.4324/9781003373094-2

the underlying issues of the strengths and weaknesses inherent in our perception. As the variables begin to accumulate – at times destructively, sometimes in more benign ways – they only serve to further the preconception that sound is this mysterious, unknowable thing.

Spoiler alert: for all the factors and variables, it is not. In taking some of the mystery out of sound, it would follow that our best course of action is to track down and identify all the data inherent in those bits and pieces, measuring what we can, quantifying what we can't, and ultimately arriving at our own, individual sound truth.

In our pursuit of sonic truth, it's best to start with the measurable criteria. Think of these as sound macro-categories as we use these four measurements in an attempt to quantify the physical properties of sound.

First, sound needs a medium; it only happens through the movement of the molecules of air around us. It emanates outward from its source, with those vibrations being transmitted through the air (or sometimes through another material, more on that later) before arriving at our ear, where that receptor feeds the information to the brain, which translates all incoming sound data into a defined and actionable set of information.

1.1.1. Frequency

Frequency is a measurement of how often a sound wave vibrates, or cycles, over time. Like any vibration, it's a repeating pattern, at least in part, and we can easily measure the number of sound wave cycles that occur per second. Frequency and its musical counterpart, pitch, are part of the same concept, but pitch can be more complex than a singular frequency measurement might allow. Figure 1.1 shows an example of a sound wave as a pure sine wave, but

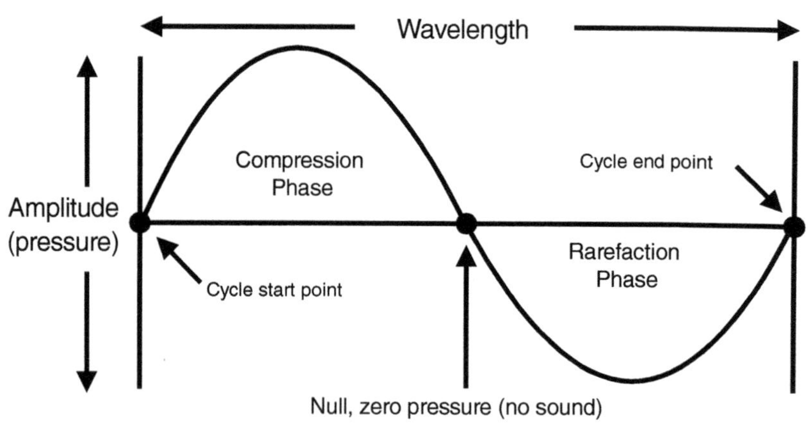

FIGURE 1.1 An example of a sound wave

with the understanding that real-world sound waveforms are rarely this pure in form; this waveform is just for illustration.

As seen in the graph, there are two distinct parts, or phases, of the sound wave, which are necessary to create one complete cycle that we can perceive as sound. The two phases are known by their actions – compression and rarefaction – which identify the process in the medium in which the sound wave is occurring, which can be air, metal, water, or anything else. In a not-entirely-accurate analogy, imagine the ripples that would appear on a pond's surface if one were to drop a pebble into the water (see Figure 1.2). Those ripples mark the disturbance initially created by the pebble, emanating outward from the source of that disturbance. The difference here is that first, sound is not a pebble and isn't a singular disturbance, but more importantly, sound emanates outward from its source in a spherical 360° pattern, not just in the two dimensions we see in the water surface example.

The first part, or phase, is called the *compression* phase. When that pebble hit the surface, it pushed a small wave of water outward as the initial disturbance caused a slight compression of the water near its point of impact. But physics dictates what happens next, and the laws of inertia are in play now – that is, for every action, there is an equal and opposite reaction – and so, the water "pushes back" as it tries to regain equilibrium and re-establish its original, undisturbed state of a flat surface. This leads to a second phase, which is that trough right after the compression phase in which the water surface reacts to the initial disturbance. This trough phase is known as the *rarefaction* phase.

Again, the graphic shows a sine wave, which is a pure tone. A sound waveform is rarely this clean in nature – sound waves more often occur as different frequency combinations interact and interfere with each other. The sum of these interactions is collectively referred to as harmonics, where different

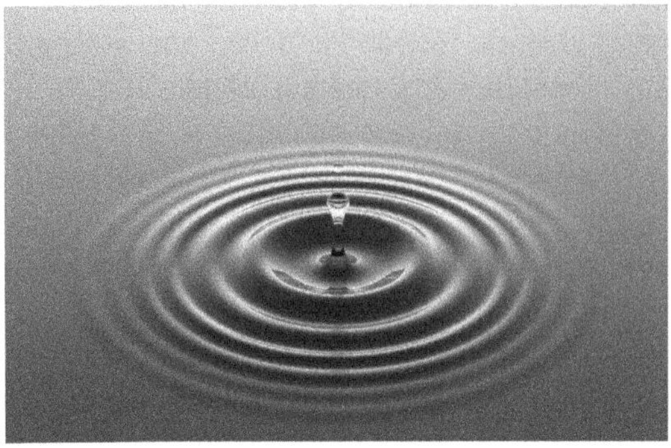

FIGURE 1.2 Ripples on a pond

frequencies layer on and interact with each other, sometimes in weird and unexpected ways, often creating the illusion of different tones, or pitches, within the original sound. For example, if one were to try to whistle this frequency as a note, it would not be as smooth as this sound wave. It would likely have some jagged edges, representing any number of other frequencies and harmonics created along the way. Despite that, one could still see the repetitive nature of each sound wave cycle as it happened.

Again, sound is more than just two dimensions, which the water surface in our example is, and this is where our pebble ripple analogy comes up short. For a more accurate representation of a sound wave, imagine those ripples happening in three-dimensional space, emanating in a spherical 360° outward from a central source, and you'll begin to have a better idea.

In our terminology, these cycles per second are identified as Hertz; also note the use of "k" as a placeholder for 1,000. For example, 1,000 Hz is often notated as 1kHz. As in physics, engineering, and math, this shorthand representation is typical in sound.

Generally, we humans can hear frequencies between 40Hz and 16,000Hz, but the entire range is most often represented as rounded off to a broader, all-encompassing scale of 20Hz–20kHz. This full scale is somewhat aspirational for us humans, even at our most sensitive best, but the 20Hz–20kHz scale is a benchmark range by which a lot of audio gear is measured. It is important to note that not many of us will hear that entire range for all those reasons mentioned earlier, of individual biology, genetics, age, and environment. From here, we have our first disclaimer that, yes, indeed, differences in perception between individuals are likely.

Everything sound can pass through, from speakers and headphones to microphones and even the cabling necessary to connect all those devices, right up to and including our ears – every part of this sound chain has a measurable range of frequency sensitivity. In the electronic sound chain, the ideal is to have gear and processes that approach the benchmark response of 20Hz–20kHz. This rounding off ensures that whatever the gear or process is that is passing audio in the chain, its ability to respond to or reproduce sound will encompass at least the totality of most humans' hearing range.

As the frequencies change, one can start to see how we humans perceive and manage these sound inputs – that some frequencies seem louder, while others disappear, such as at the ends of the spectrum. This is human perception at work, and we begin to see that not all frequencies are created equal.

(See *Media 1.1m – narrated spectrum sweep*.)

Figure 1.3 gives you a more relatable idea of the amplitude or loudness ranges for each type of sound, which leads us neatly into the next section.

Relative Loudnesses

Decibels (dB)	Source
150	Rifle fire, close
140	Jet engine, 10m away
130	Threshold of pain
120	Rock concert
110	Thunder, close
100	Shouting, close
90	Heavy traffic, rush hour
80	Noisy restaurant, vacuum cleaner
70	Dishwasher, washing machine
60	Normal conversation
50	Moderate rain, refrigerator
40	A quiet office, light rain
30	Quiet room, residential
20	Whispered speech, 5–10 feet
10	Breathing (normal, internal)
0	Threshold of hearing, 6–10dB

FIGURE 1.3 Amplitude ranges for different sounds

1.1.2. Amplitude

This is the vertical axis of any graphic representation of sound. Amplitude is measured in decibels – a logarithmic scale (using a base of 10). Ten times the logarithm to the base 10 of the ratio between the powers of 2 signals: dB = 10 log (P1/P0). P0 is a reference power value with which another power value, P1, is compared.

I know. I promised to go light on the math. Still, it is also important to get across how complicated the science of measuring sound levels can be … and we haven't even considered the human perception component yet.

The decibel – or, more correctly, deciBel – and its abbreviation, dB, come to us by way of its creator, Alexander Graham Bell himself, who is represented by the Bel at the end. We employ the "deci" prefix to measure in increments of 1/10th of a Bel. The Bel was originally a measure of the amount of power a signal lost in level over a 1-mile distance of telephone wire, but this scale turns out to be too big and unwieldy for the sensitivity of the audio requirement; the level difference is too great to work with and doesn't provide enough definition. Therefore, the Bel has been divided into tenths to make it easier to work with while, at the same time, not much more accessible to wrap one's mind around!

When quantifying acoustic sound pressure – that is, the level of a sound wave in a physical environment – we measure it in terms of sound pressure level (SPL), or, more specifically, dB SPL, to clarify that there are matters of atmospheric pressure at work on a sound wave. We will also see dB applied to measure a system's capabilities, but when used to measure the performance or the amplitude range capabilities of audio gear, it's the straight "dB" value.

While 0 dB might work as the theoretical start of a scale of measurement for a device or system, it does not exist here in the real, practical world. We do have a relative threshold of hearing, becoming more apparent when the environment gets quieter – this is approaching practical silence for us humans. Ultimately, there is no such thing as pure, total silence in the physical world unless in the vacuum of space, an idea that itself goes back to the theoretical domain because there's no way humans could exist in that environment long enough to experience it. There's a point in the scale at which humans can begin to perceive sound – it's a very low measurement in terms of decibels, but it's also not zero. On the other end of the scale, we arrive at a loudness measurement where sound waves hurt and cause actual physical damage to our hearing mechanisms. The threshold of physical feeling and discomfort (generally considered to be 120 dB) gives way to the threshold of pain not long after – usually considered above 130 dB.

Humans can perceive those different steps of loudness at a ratio of 1:10,000,000 or more, making it readily apparent why we stick with the logarithmic scale. Measuring loudness would otherwise be too broad a measurement and need many more steps of definition over the entire scale.

While we might need to measure the amplitude capacity of a piece of audio, as in a microphone that can handle up to 120 dB of sound pressure level (SPL), loudness is also a relative state within any given environment, program, or system. This difference between two (or more) loudnesses is called *dynamic range* and will come into play later.

Sidebar – Hearing Loss

We're surrounded by sound stimuli that interfere with and influence our hearing response, some of them intense enough to even affect our hearing over the longer term – we are bombarded with sound and noise regularly and at all times of the day or night. All that stimulus weighs on us and will degrade our hearing mechanisms over time, so we arrive at yet another example of "individual performance may vary." For example, a person who spent their life in the country is likely to have a better overall hearing response than someone who has lived in an urban setting. One individual was simply living in a louder environment than the other, but that constant input had a cumulative effect. How we hear is directly affected by what we hear, the sounds we are subjected to, and how often.

We live in a world of constant noise. While the noise of city sounds, wind, water, and vehicles might not be something we consciously notice as we go about our day, it is ultimately insipid and corrosive to our overall hearing sense. It will eventually wear our listening abilities down to a place of diminished hearing capability, sometimes to the point of becoming an impairment.

It is insidious – an action as simple as having the car window rolled down at highway speed will affect our hearing abilities; the combination of road noise and the roar of wind creates noise at great enough amplitudes to cause significant, cumulative hearing loss over time. This is just one small example: there are so many other noisemakers in daily life – air conditioning systems, industrial and other machinery, airplanes, traffic, commuting on public transportation, construction noise – the list goes on. The simple act of listening to a TV or music at a consistently loud amplitude can harm overall hearing sensitivity over a single day. This sometimes causes a temporary shift in hearing sensitivity, with earbuds and headphones being the worst offenders.

For example, after attending a loud show, one might experience a ringing sensation in the ears. Typically, it's temporary, and the effect goes away with some rest and relative silence. However, one doesn't need to attend rock concerts to experience this phenomenon; it can also happen over the course of a relatively normal day as well. As evidence of this, when turning on a TV or radio for the first time in the morning or putting earbuds in with the volume set at the same level as they were the night before, they will often seem to be at a much greater volume than we remember from the night before, illustrating the point of how hearing sensitivity shifted during the previous day.

The point is that this degradation is at work in our day-to-day lives, and exceptional events aren't necessary to do long-term damage. This phenomenon is called a temporal shift, characterized as a temporary loss of hearing sensitivity brought on by the cumulative effects of the amplitude of the sound environment we might have experienced over time. This is evidenced – by our hearing recovery the next day – as the loss of sensitivity over the course of hours instead of months or years. The shift occurs because hearing sensitivity gradually diminishes over time from exposure to sound stimuli in a particular environment, so we often turn our sound sources up louder to compensate. In the short term, we're not always aware of how that volume setting might have crept up on us as we went about the day, but living in the city and dealing with all its sounds – along with maybe just wanting to hear that music a little louder – will all be factors in how and why it occurs. In making our devices louder, we're only trying to overcome a noisy environment to hear the programming more accurately.

That loss of sensitivity is not always permanent, but repeated temporary losses of sensitivity over time begin to add up, and accumulated exposures can ultimately lead to a more permanent loss of hearing sensitivity. It is important to note that hearing loss of this type is not just for factory workers, airport personnel, and other occupations we might equate with having loud sound environments. Sadly, a lot of hearing loss is preventable because it's self-inflicted with personal listening choices. Here, the daily temporal shift might accelerate hearing loss over even less time as listening levels are adjusted upward more rapidly over the course of a day. That driving, upbeat music chosen for a morning workout routine or commute could be the largest part of the problem and likely why a significant portion of the population is experiencing hearing loss, in greater numbers than ever.

It's a straightforward concept – hearing loss often occurs with repeated and prolonged exposure to damaging sound pressure levels. The mechanics of the ear indicate that hearing loss usually begins in the higher frequencies first. See Table 1.1 for damage risk criteria, per the

TABLE 1.1 NIOSH and OSHA noise tolerances

NIOSH Standard		*OSHA Standard*	
Sound level (dBA)	*Duration (HH:MM:SS)*	*Sound level (dBA)*	*Duration (HH:MM:SS)2*
82	16:00:00	85	16:00:00
85	8:00:00	90	8:00:00
88	4:00:00	95	4:00:00
91	2:00:00	100	2:00:00
94	1:00:00	105	1:00:00
97	0:30:00	110	0:30:00
100	0:15:00	115	0:15:00
103	0:07:30	120	0:07:30
106	0:03:45	125	0:03:45
109	0:01:53	130	0:01:53
112	0:00:56	135	0:00:56
115	0:00:56	140	0:00:28
118	0:00:14	145	0:00:14
121	0:00:07	150	0:00:07
124	0:00:03	155	0:00:03
127	0:00:01	160	0:00:01

National Institute for Occupational Safety and Health (NIOSH) and the Occupational Safety and Health Administration (OSHA).

Here we see the differences between the NIOSH and OSHA Standards. The two governing bodies issue their own recommendations for exposure time at each decibel level before the risk of hearing damage and/or loss. For example, according to OSHA, a person can remain in an environment with sound pressure levels of 95 dBA for up to four hours – dBA indicates a measurement scale more closely calibrated to Human Hearing Response (HHR). NIOSH says that a person is only safe in that same environment for an hour or less.

Another effect of repeated exposures to loud sound is tinnitus, with the primary symptom being a "ringing" sound or tone in the ears, often strongest between 3kHz and 6kHz. While it is not a loss of hearing, it can get in the way of daily life in more advanced cases, interfering with our ability to concentrate, sleep, or even hear certain sounds accurately. At its worst, it can compromise our ability to understand conversational speech. There is no proven cure yet, although a burgeoning field of remedies and therapies touting relief from those effects exists. These are almost entirely anecdotal and, for the most part, clinically unproven – indeed, more study is needed in this area.

The best course of action is intervention and taking whatever steps are needed to keep that damage from occurring in the first place. The best course of action is preventative, and by lowering the volume of personal listening systems and protecting our hearing when in loud sound environments (per the OSHA and NIOSH charts previously), we can hope to retain our hearing capabilities for longer in life. Controlling listening levels in the environments where we work as sound professionals will also go a long way toward preserving our ability to hear.

1.1.3. Frequency Response

Assessing and measuring Human Hearing Response (HHR) is one application of frequency response, but the term has much broader applications and implications for us in the sound field. Indeed, humans have a measurable frequency response as well. From Figure 1.4, we can see that hearing capabilities vary by frequency content; in some areas, they are more robust, and in others, they are weaker. Human hearing is best in the middle frequencies (generally) from 400 Hz to about 3,000 Hz. Conversely, there is a lack of sensitivity in the upper and lower extremes of the range – we don't hear lower bass frequencies and higher frequencies as well as the midrange frequencies. Moreover, as discussed in the

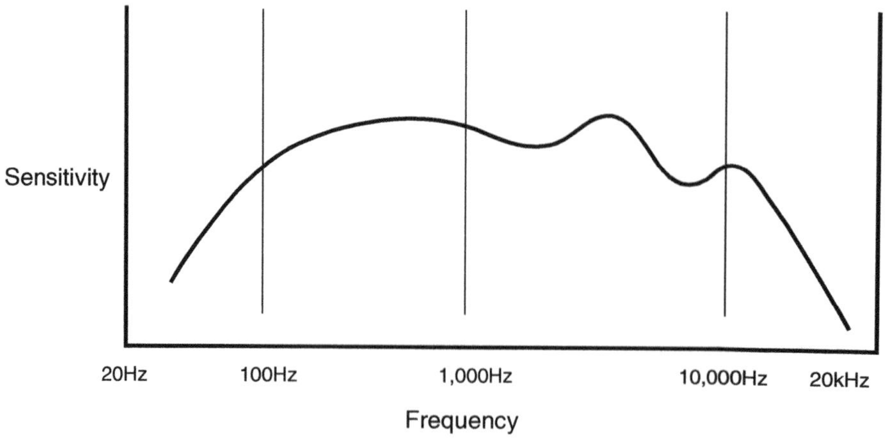

FIGURE 1.4 Human Hearing Response (HHR)

sidebar earlier, the higher ranges of the frequency spectrum are fraught with individual response capabilities and deficiencies and degradation effects due to time and exposure.

As you can see, human hearing falls along an imperfect curve, compounded by the fact that this graph is a generalized response measurement meant to represent most humans. Many individuals may not be able to hear this well, whereas others could have an even better frequency response than this. As the variables start to reveal themselves, this generalized frequency response curve represents a consensus of the population.

Therefore, the term is broader, going from the person-centric HHR to the statistically inclusive Frequency Response, to help better measure the equipment and systems needed to manage, store, and handle sound. Consequently, every device, circuit, and bit of wiring that audio travels through also has a measurable frequency response. Perfection is the common denominator goal here; however, it's not practically achievable, so we strive for the best frequency response possible to achieve the most uniform sound experience possible. It wouldn't be possible (or practical) to create individual audio programs ideally suited to everyone's hearing response; the scope of that encompasses far too many variables. Instead, the goal is to strive for the best overall frequency response technically possible across all components of the sound chain.

Figure 1.5 shows a graphic display of the frequency response for a Neumann U87ai microphone. Again, in most cases, the goal is to reach a frequency response that is as "flat" as possible – without the hills and valleys of a less-than-perfect reaction in specific frequencies. That ideal itself is unrealistic, however, and as the further realities of physics and acoustics come into play, there won't be enough ways to counteract all the factors to create the perfect microphone.

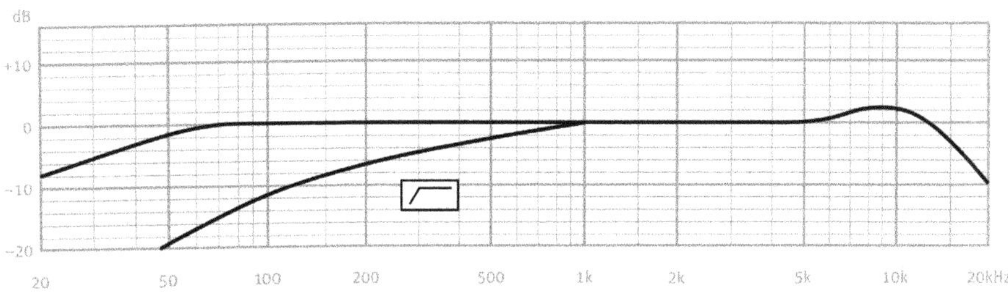

FIGURE 1.5 Frequency response of a Neumann U87ai microphone

Source: Photo courtesy Georg Neumann GmbH.

In this example, the microphone's measurement has some variation in its response, but overall, the variations are minor in the frequencies that matter most to us humans. The frequency response of this microphone is best suited to capturing the nuances of the human voice, typically strongest in the frequencies between 100 and 5,000 Hz, as well as many acoustic instruments. But this brings up another point – that many audio devices, including microphones, are precision tools and very often application-specific. Finding a microphone with a flat response along that X-axis is nearly impossible, and the microphone only gets more expensive the closer it gets to this. Designing a microphone that works better for a specific task is a much more realistic objective.

As a result, engineers and recording studios invariably end up with a closet full of microphones, each designed and optimized for a particular task. There are voice and vocal microphones, drum and instrument microphones, microphones made for ambient and environmental recordings, and so forth. Accumulating a locker full of microphones might seem excessive and redundant, and it can certainly get expensive, but they're just tools in a toolbox. Imagine how incomplete a toolbox would be if it only contained a couple of hammers, or maybe just three wrenches the same size.

To arrive at an accurate frequency response, several tools can help measure how a piece of equipment or an entire system reproduces or responds to the entirety of the frequency spectrum. In calculating the frequency response of that microphone – or any system – we can apply an audio spectrum analysis at the output of that device. At the same time, we play certain frequencies through the signal chain or the specific device in question. The assessment tool for this is white noise, defined as all frequencies happening simultaneously and at equal amplitude (see Figure 1.6).

(See *Media 1.2m – white noise.*)

With this new concept, the frequency response concept can be applied to the entire range of devices and systems that pass audio, from speakers,

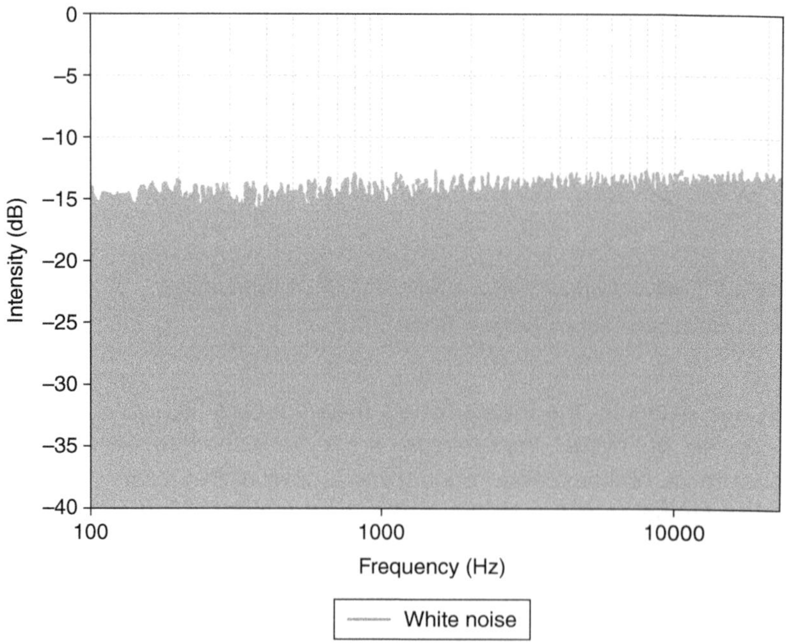

FIGURE 1.6 White noise spectrograph

headphones, microphones, microphone preamplifiers, and consoles, all the way down to the circuitry that connects them. Each component will have a measurable frequency response, and together with this new definition of audio fidelity, we can make sure everything has the best (read: flattest) frequency response possible for the task it is designed to do.

While listening to white noise at higher amplitudes is undoubtedly not a pleasant listening experience, it is a valuable tool for measuring a system's capabilities (or weaknesses) when reproducing a given sound wave. Note how it sounds hissy and unpleasant, a direct indication of the disparities in our very human inability to hear all frequencies equally and how people cannot perceive all frequencies accurately and with the same weight. The human ear–brain hearing "system" is composed of its receptors (the ears) and the processor stuck between them – the brain – and is at its most discerning in the midrange frequencies, generally between 1,500 and 5,000Hz. Hence the scratchy, unpleasant sound.

To get around this, another noise form can be employed to better compensate for the inadequacies inherent in human hearing abilities. The ear–brain system, as one might have guessed by this point, is not as simple a system as it would seem, nor is it impartial and objective – instead, it's subject to all of the perceptions and tendencies that make us human, from genetics to

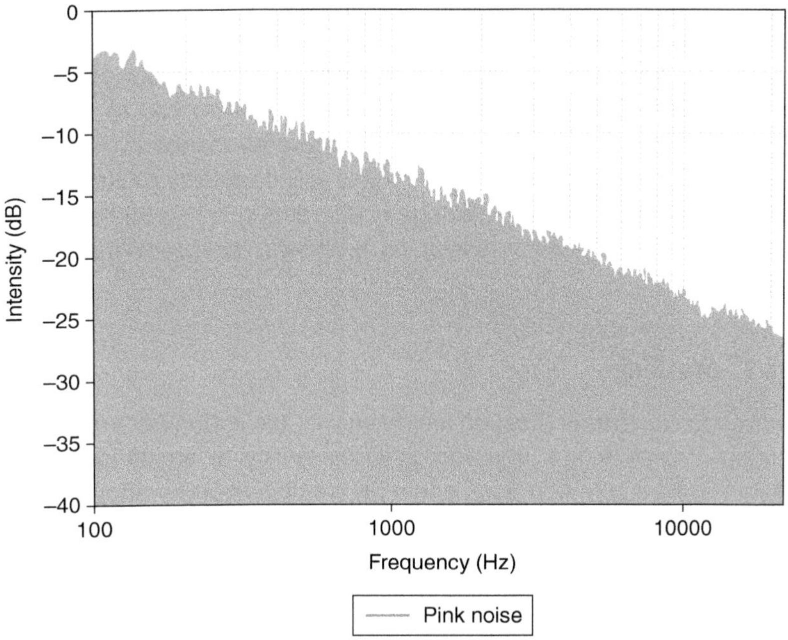

FIGURE 1.7 Pink noise spectrograph

environment, on down to and including how rested we are, and even how we might be feeling on any particular day. We can try our best to compensate for these differences, but the fact remains that we are still imperfect listening devices, even at our best.

Engineers will employ a sound called pink noise (see Figure 1.7) for the measurement of acoustic sound material and for its tailoring to the human hearing system – that is, simulating it as experienced in the world and playing into the environment.

(See *Media 1.3m – pink noise.*)

While it is still unpleasant to listen to, this sound waveform is compensated in some general ways to account for the inadequacies of HHR, primarily by increasing the amplitude of the bass (low) frequencies, attenuating the midrange, and accentuating higher frequencies to some extent. As mentioned, this tool is applied in the acoustic or real-world realm, making it different from the white noise typically used as an equipment measurement tool. Pink noise is designed to better align with human reception in real-world acoustic environments.

One application would be a venue where sound is to be played back or amplified, such as in a concert hall, where engineers apply this compensated noise to measure the actual acoustic response of the space itself. It is applied at the approximate amplitude of the program material to be played in that space. It gives the best analysis in measuring how the space will respond to the acoustic energy that will be heard there later. That noise signature is then used to measure how that space affects and influences sound reproduction down to individual frequencies. The information gathered is then used to adjust some frequency areas to help compensate for the overall response of the space. Again, the objective is to control the room's response and make it as flat as possible, but with the added challenge of addressing and reducing unwanted acoustic phenomena, such as reflections, feedback, and overall inaccurate frequency response.

1.1.3.1. Masking

Masking occurs when one sound interferes with the intelligibility of another by obscuring or overtaking it, through either frequency or amplitude (loudness) interference. It's a common issue and a primary focus of our efforts in making a coherent, listenable soundtrack.

First, we have *frequency masking*, which occurs when some frequencies interfere with and obscure the frequencies crucial to understanding the material we need to hear. Examples would be trying to hear a dialogue track with strong background noise, such as air handling, rushing water, wind, traffic, and so forth.

In short, sounds with greater midrange intensity effectively mask the dialogue we need to hear. Because of human sensitivity to this range, this type of noise more easily confuses and distracts our focus. The worst offenders are sounds with a broad range of frequencies present, as with white noise, and many of the most disruptive sounds tend to be broadband ambient or environmental noise. Because these types of noises have so many frequencies represented – as with white or pink noise – they are much more likely to obscure the frequencies we need to hear to understand the sounds being conveyed.

This is a constant issue on a film or video set, particularly when the location is not entirely under the control of the production. It's a problem that can present itself in any setting where the recording of clean voice or dialogue is necessary. The in-home setting of a podcast might have some air conditioning or other air-handling noise, for example, whereas a larger film set might have many other environmental sounds that need to be addressed. The list is endless and includes the surf, wind (as mentioned earlier), and traffic noise that might come with the physical location. In many cases, the production might have little to no control over the surrounding environment when making on-location production happen. A big film's budget might allow it to "own" the actual street they're shooting on and shut down the traffic flow that would otherwise interfere with getting good sound. Still, there often isn't that level of control over those elements when they happen just beyond the range of the location that has been secured. Wind and weather, people going about their day, birds and other animals, planes, traffic, sirens, and trains – all are generally resistant to any efforts to control them, to say the least. So, the Sound Department must find ways to contend with those noise events as they happen.

Microphone choices, set location, and exercising control over noisemakers will all come into play, as will the audio processing tools that can be applied in post-production. While the best strategy is to shut down all unwanted noise at the outset, the reality is that it's often impractical to do, and most noisemakers will continue to defy any efforts at taming them. A mixer and their crew can bring all their powers and skills to bear, but many situations will still require other means to mitigate those issues after the fact.

Later in the chapter, we'll discuss noise reduction tools and other means of dealing with noise issues and sound repair, leading to the mindset of many aspiring story producers: "Why bother? I can just fix it in post!"

Well, first off – that faith in technology is misplaced. And besides, people working in post-production do get tired of hearing that one; it's just kicking the can down the road and making it someone else's problem in post-production. The simple fact is that it will always be better (and ultimately, easier) to get it right the first time. Indeed, there is a world of tools and techniques to correct those issues and make those things better in post, but there are also limitations to what it can effectively do. That time and energy will always be better spent on the more creative aspects of soundtrack design than the need to have a viable, listenable dialogue track. Technology can only go so far, and even the best tools aren't going to salvage a track that was simply poorly recorded.

Temporal masking is simple in concept and often the easiest to mitigate in the moment, at least while everyone is still on set. Temporal masking is when an unwanted noise coincides with and obscures a quieter sound – often the dialogue we're trying to capture and record cleanly. It can be as simple as a door slam coinciding with (and consequently obscuring) a critical bit of dialogue, but it could be any overlapping noise getting in the way of the sound we need to hear. Possible noise sources could be the sound a simple prop in a scene might make, such as glassware clinking or mechanical tools, but it also encompasses all manner of hits, taps, noisy props, and even the clapping of hands or footsteps. Any sound that is momentarily louder than the dialogue we want to hear will interfere with the intelligibility of the dialogue and potentially ruin an otherwise clean sound recording.

As stated, these types of errant noise are always better eliminated at the source, and we can (and absolutely should) try to head them off at this point in the process. A saving grace might be embedded in the process of narrative film production – throughout production, many takes will be shot, allowing more opportunity to take meaningful measures to ensure noise doesn't interfere with any subsequent takes.

Temporal masking presents a more significant problem when the production is more reality-based, such as in a documentary or an interview situation. Asking interview subjects to repeat themselves for any reason, such as being interrupted by sudden and unwanted loud noises, is likely to take them out of the moment and disturb the flow and continuity of the information they're trying to convey. In that situation, the sound crew will go to great lengths to anticipate and remove all potential noisemakers by silencing systems and quieting or removing any potentially noisy items before they become an issue.

Sounds simple on paper.

This type of masking can be significantly more challenging to deal with in post-production, depending on the nature of the noise, so anticipation is critical. Experienced mixers on set develop a sixth sense for these issues and noisemaking items by treating and removing them before they become a problem. Any props handling or wardrobe that might make noise and interfere with a line of dialogue is identified and dealt with before any further takes are shot. Post-production tools are limited in this area, mainly where the offending sound might be greater in amplitude than the sound (again, mostly dialogue) we're trying to keep.

Sometimes, we'll see more seasoned actors handle the issue themselves; for example, during a scene where the actor is supposed to applaud during dialogue, a more experienced actor might instinctively know to "pull" their handclaps – that is, to make it look like they are actively clapping their hands, but not let their hands actually come together to create a sound. An actor getting dressed and putting on a jacket might pause on using the zipper, knowing that it might make an unremovable noise over their line if they were to zip it up while speaking, especially if their lavaliere mic is on the sternum, as is typically best practice. These actors benefit from their experience – they've been through it all before, and one can guess they have likely spent enough time in the booth replacing lines that otherwise would have been perfectly fine, and they want to avoid that process if possible.

In any case, a grateful Sound Supervisor and Dialogue Editor on the other end will benefit from the forethought that went into making their tasks more manageable. That person might even be you.

1.1.4. Velocity

There's sometimes a question about why we need to worry about the speed of sound. On the face of it, the velocity of a sound wave would seem to have little to do with any kind of visual storytelling, as it can't be controlled and (probably) won't come into play on set. There are some physical realities, however, and some storytelling conventions that could be factors here. But first, let's define the speed of sound.

Like so much of what goes into our hearing response, the velocity of sound itself is variable, dependent on and affected by environmental conditions like temperature, humidity, and altitude. While it's a variable in real life, we often default to a baseline measurement of 1,130 ft/sec, which is, again, entirely contingent on the environment and the medium; 1,130 feet per second occurs at sea level, at 70° Fahrenheit, and at 70% humidity. It follows, then, that velocity varies in cold or high temperatures – increasing slightly as the air warms, slowing as it cools. Warmer, denser air molecules mean faster-traveling sound waves, whereas colder air that is less dense results in slower-moving sound waves. Atmospheric pressure can also affect sound velocity – atmospheric pressure decreases with altitude with fewer molecules being available, resulting in slightly slower sound. Sound at sea level and below means greater atmospheric density and, therefore, slightly faster sound.

The medium through which sound travels also plays a significant role in the speed of sound. We humans experience most of our sound in the atmosphere – that is, through the air around us – but sound waves also travel through other solid and semi-solid substances as well. For reference, sound travels more than four times faster in water at 4,800 feet per second, at 11,700 feet per second in wood, and at 18,000 feet per second in steel – demonstrating again that the denser the medium, the faster the sound. This makes the point that the greater the density of any atmospheric conditions – mainly higher humidity and lower altitudes – the higher the velocity of sound will be.

This comes around again to why, in our world of sound design, we need to consider the velocity of sound in the first place. Microphones are close enough to any sound sources being recorded that any time differential between the visual and its corresponding sound would be virtually unmeasurable; the velocity of sound is not a practical concern for us when on set.

Surprisingly, tradition and convention have a lot to do with it, and the main area we have to deal with it is in post-production.

First, note that sound and light do not travel at the same velocity. Light travels approximately 186,000 *miles* per second, whereas sound (in our atmosphere) moves at a much slower velocity of 1,130 *feet* per second. That's a huge gap; the two are on vastly different scales of reference – we're measuring on a scale of miles for light, versus only feet for sound. The difference in physics between the two cannot be reconciled; it will require a variety of workarounds, a little compromise, and, sometimes, a bit of creative thinking.

One area where difference in velocity reveals itself to us is on the distribution end of the process and in the size of the space where we might be experiencing our visual story. This won't be a significant factor in many instances where picture is being projected with sound. For example, with a typical television screen viewing setup in a room in a house, there isn't a great enough distance for us to notice any timing difference. Sync issues quickly become noticeable, however, as the viewing space becomes larger; there is a point where those differences become significant enough to be perceptible to most people.

We'll start with a more extreme example first.

Imagine a large concert hall that measures 150 feet in depth. While such a space would be ideal as a grand music venue, it's on the large side for modern cinematic theater purposes and would likely be considered too big for film projection. Halls of this size and larger have been used for hundreds of years as performance spaces; however, audiences have tolerated the delay.

That delay can become significant with distance from the stage – assuming a sound velocity of approximately 1,130 feet per second (close enough for illustrative purposes here), it will take approximately 0.133 of a second for the sound to travel the full depth of the hall (see Figure 1.8). While such a small fraction of a second doesn't seem a significant measure of time, it becomes significant if we look at that delay in terms of frames per second. Assume a projection rate of 24 frames per second, which is on the order of .042 of a second per frame. That implies that, in this case, people at the back of the venue will receive their sound over three frames late.

This is enough of a delay that many people will notice, particularly on what we'd call "hard sync" actions like door slams, hits, gunshots, and so on. The sound would be delayed enough that it would prove distracting to some. The bottom line is that the delay becomes more noticeable as distances increase, as the sound wave takes more time to traverse greater distances.

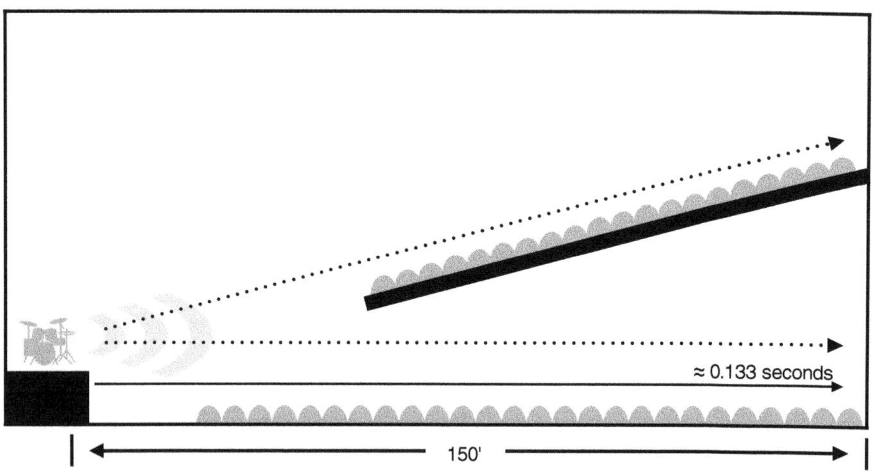

FIGURE 1.8 Here we see the amount of time it takes for sound to reach the back row in a large venue. Sound takes around 0.133 seconds to reach the back row, arriving more than 3 frames late, making a film at this distance appear to be out of sync.

The physics of this situation can't be altered. The phenomenon is a given; it is simply accepted as truth, and short of supplying our audience with headphones (which would completely defeat the whole point of hearing music in a live venue), there isn't a technology that propels sound forward at greater velocity. The best that can be done is to devise a less noticeable compromise.

In the case of the larger film theater, the solution is to slightly advance the film's soundtrack, usually on the order of half the delay as measured at the back of the space – enough so that the middle of the theater space ends up exactly in sync. The compromise is that any issues caused by the physics of the situation have only been dispersed, making them slightly less noticeable overall. The front of the theater will get its sound slightly early, with the sound reaching the back of the theater now only slightly delayed from the visual – all with the hope of making that delay less noticeable.

Modern theater design further helps compensate for this with an architectural solution, through stacked seating and a steeper incline from front to back (see Figure 1.9). The idea is to push all the seats higher and closer to the screen and speakers, helping to reduce the distance from the screen to the most distant seating. This also improves sightlines, but the main advantage is to decrease the time lag for the back of the space by reducing the distance between the screen and the seats.

Another way to account for the velocity of sound is within the actual soundtrack design of the story itself and how the typical viewer would perceive that delayed sound. The norm is that in film, sound and light need to travel at the

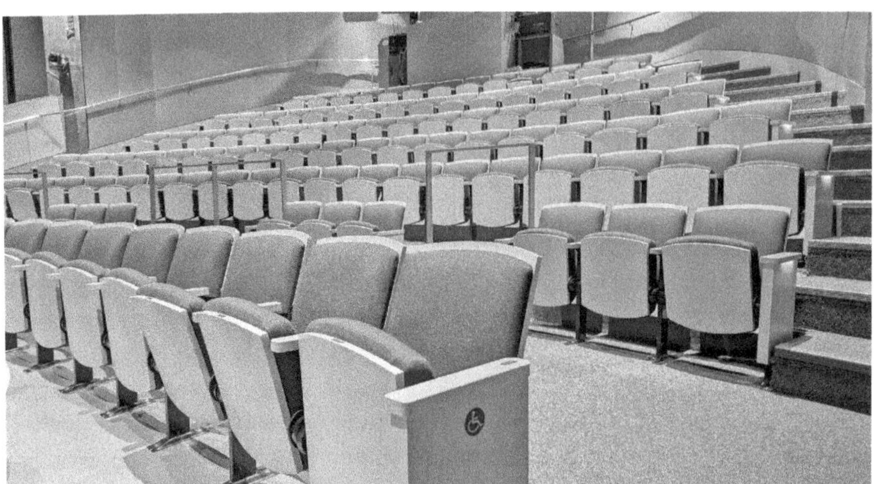

FIGURE 1.9 Modern theater design incorporates steeper angles for seating to bring audiences closer to the screen

Source: Courtesy WGBH Educational Foundation.

same speed. As with any other rule in the world of storytelling, this convention might have been made to be broken, but the goal is to keep sound as synchronous as possible so it doesn't distract the viewer. My own hard-won lesson came early in my career.

When working in a one-stop post house that catered mainly to advertising clients, a colleague and I took on the sound design for an aspiring filmmaker when we were all just getting started on our path. The film was an angsty romance, but (spoiler alert) the star-crossed couple ends up back together. In the last scene, they kiss on the Massachusetts Avenue bridge between Boston and Cambridge at night and during Boston's Fourth of July celebration. Our intrepid filmmaker pulled off the shot during the July 4th celebration that year. This was no small feat, as Boston's fireworks display is a big deal that attracts thousands to the Charles River that the bridge crosses. In some ways, it was an ideal setup, as the bridge was closed to traffic, and the fireworks display over their heads was the perfect backdrop for the whole scene. Miraculously, the mixer managed to record a dialogue track that was mostly clear of the actual fireworks in their takes.

Later, during our work in audio post, we decided to be as factually accurate as possible. We determined the distance of the fireworks display from the actors and the time delay they would have experienced from their position on the bridge. We came up with a delay time of 1.7 seconds to their position on the bridge, considering the ground distance to the fireworks barge in the river, along the hypotenuse to the approximate altitude of the fireworks when they exploded.

And so, we built a track with the fireworks' whistles, explosions, and reports, first placing the fireworks sounds exactly in sync, then delaying that entire effects track by the 1.7 seconds we had determined to be the actual delay time.

We were proud of ourselves for figuring this out and couldn't wait to screen the scene for the director.

The scene only played for a few seconds when he made us stop and asked us what we thought we were doing.

We started by explaining our math – again, we were very proud of that hypotenuse thing – and he cut us off, saying that, all calculations aside, it just looked wrong. We argued, discussed, and ultimately had a tough time letting go of the preconception that it should be entirely physical-world accurate. We finally came around and let it go because, in the end, the director's intuition was correct. The scene didn't work the way we had designed it, particularly at the end when the camera tilted up to capture some of the final explosions before fading to black. No matter how hard we tried to find the rationale, making the sound for those explosions happen accurately and so much later than the visual just looked like a mistake.

Here we are, then, arriving at a long-established convention that, in the world of film and visual storytelling, light and sound generally need to travel at the same velocity unless there is a good reason for them to be more realistic. As said before, there will be exceptions to this rule – thunder sounds happening exactly in sync with the lightning visual can become tedious over the course of a scene, so variations in sync might help to heighten the drama.

A great example of an exception to our rule occurs during the sniper scene in *The Hurt Locker* (Figure 1.10), where we see Ralph Fiennes' character get shot by a distant sniper. We don't hear the actual report of the gun until a time noticeably later, and then at a much lower volume than the guns being fired nearby, making it easy to miss. This plays on the confusion of the moment, with the distant sniper shot juxtaposed against the close-up chaos of the scene. The soldiers start firing somewhat indiscriminately until they narrow it down to a small building, with the sniper being perhaps a thousand yards away.

With some basic math, that would equate to an approximately three-second delay between the visual of the gun firing and the sound of that gun's report reaching us. There's also the length of time it would take the bullet to traverse that distance, so we can make some general assumptions there as well – the muzzle velocity of a Barrett sniper rifle is approximately 6,500 feet per second or nearly six times the speed of sound. Therefore, that bullet could cover the distance in about half a second, while its sound would be heard over 2 seconds after the bullet hits.

Ray Beckett and Paul N. J. Ottoson took all of this into account in their approach to the scene, making the speed of sound a component of the story.

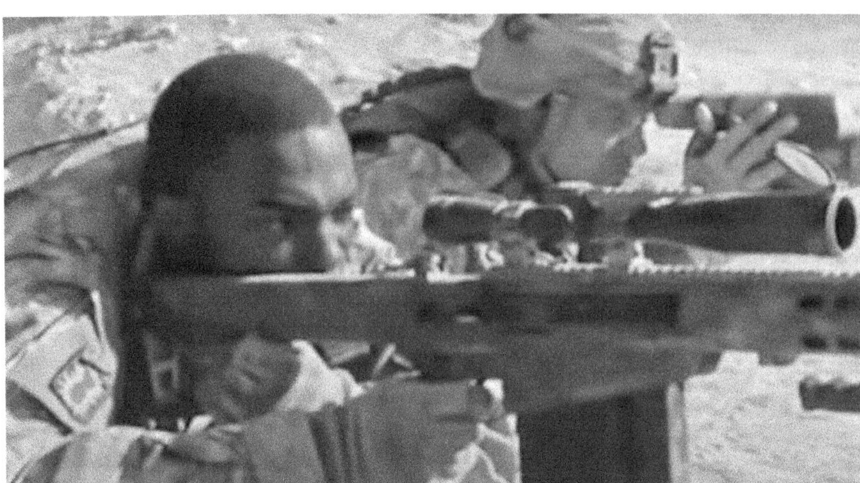

FIGURE 1.10 Sniper scene from *The Hurt Locker*

Source: *The Hurt Locker*, 2008.

Ultimately, it was effective, and that time delay became central to that scene. The members of the platoon are disoriented as an unseen enemy is attacking them; locating the source of that threat becomes a matter of their survival, and the delay in time between the bullet and its corresponding sound arriving is their biggest hurdle to figuring out where that sniper is.

1.1.5. Wavelength

At first glance, the measurement of wavelength appears to be another element that, on the surface, is less important than frequency and amplitude. However, gaining a greater understanding of all aspects and characteristics of a sound wave will go a long way toward anticipating how sound will behave, as well as heading off issues that might arise later.

In the context of acoustic sound, wavelength is the physical measurement of the distance a sound wave travels over one complete cycle – something that would seem to defy that earlier description of the amorphous nature of sound. On the face of it, something that cannot be grasped or examined in one's hand seems unlikely to have an actual, physical measurement. Still, some deeper thought would bring us to a logical conclusion that if we can now measure the velocity of sound in the atmosphere, then it must also be taking up some dimension of the physical realm. It only makes sense that a physical manifestation or measurement would corroborate that.

The wavelength measurement will also vary because it relies on that velocity measurement, which is itself subject to environmental conditions. From there, we can now derive a measurement for wavelength, which we can calculate with a simple equation: velocity divided by frequency.

Bass, or lower frequencies are less directional than higher frequencies, and as such, they are also the most difficult to contain and manage. Therefore, knowing a low-frequency sound wave's actual wavelength measurement is critical when designing systems and spaces to accurately reproduce those lower frequencies and predict how those sound waves will behave.

As one example, Hollywood mixing stages strive for the "40 Hz room" – meaning that the room should be large enough for a 40 Hz wave to fully form within that space without interference from walls and other fixtures. These factors can affect our ability to experience lower frequencies fully and accurately.

That math, then – solved for a 40 Hz wave, we use the formula:

$$W = V \div F$$

Again, the velocity of a sound wave is variable, but here we'll assume an average velocity (V) of 1,130 ft/sec. When the frequency (F) is 40 Hz, it follows that 1,130 ft/sec ÷ 40 Hz = 28.25'.

This means only that a 40 Hz sound wave requires at least 28.25 feet of physical space to form one complete cycle of the sound wave. The science of acoustics and acoustical design is much more complicated than just that simple equation, however; there are several other dimensions and aspects to account for beyond just the physical depth and dimensions of the space. However, one begins to understand how wavelength and frequency are related – the lower the frequency, the greater (or longer) the physical wavelength.

1.2. ACOUSTICS

The science and study of acoustics is a fascinating field and will only be lightly covered in this text. However, one needs to understand several fundamental factors and aspects, as they will have a discernible effect on how we do our work in sound design. Some aspects can be exploited for creative gain, whereas others need to be addressed via mitigations for potential problems and issues that might arise.

1.2.1. The Mechanics of Sound – Phase

As with the earlier discussion of phase within the context of a sound wave, there are ways in which it affects our sound signal. First, there is electrical phase, which is when two wavelengths are in direct opposition to each other, with one sound wave in compression at the same time the opposing sound is in its rarefaction phase – and vice versa. In this example, if both sound waveforms are identical, then the net result will be no sound at all when the two signals are combined; the waves will cancel each other out fully (see Figure 1.11). This can happen within the realm of circuitry and audio processing, but it does not happen here in the real or acoustic world.

While usually detrimental, it can also be a benefit when applied differently. A great example of this concept being applied for constructive use is in the case of noise-canceling headphones (see Figure 1.12).

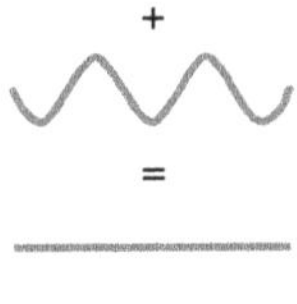

FIGURE 1.11 Equivalent waveforms 180° out of phase with each other

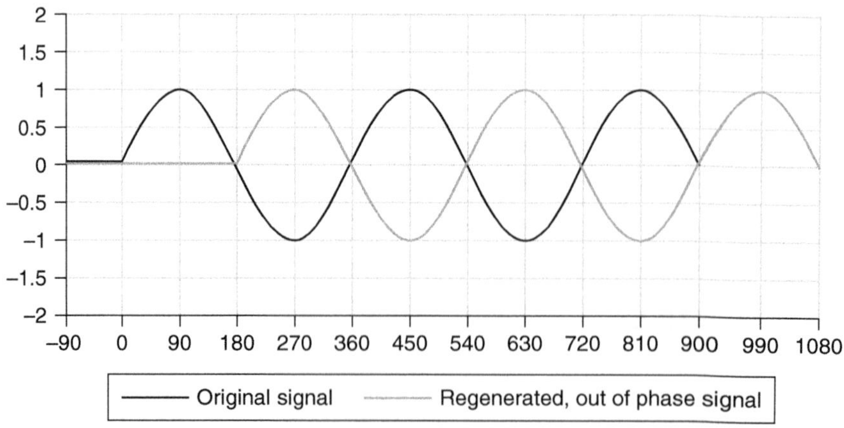

FIGURE 1.12 An example of the process inside noise-canceling headphones. Exterior noise is regenerated out of phase with itself, then played into the headphones. The result is a cancellation of ambient noise

Here, we're taking an entire sound program – in this case, the entire noise signature of the world around us – and inverting its phase with itself. We're canceling an entire world of sound instead of a singular sound waveform. Noise-canceling technology takes the noise of the exterior world and inverts it, combining both signals – the original and then a 180° inverted version of it – inside the earpieces, resulting in your ears experiencing that noise at a significantly reduced level.

However, issues of acoustical phase are more complex than electrical phase due to how sound behaves in our physical world, where reflections and other sound behaviors can prevent the complete cancellation of two sound waves. Instead, certain frequencies within a complex waveform, such as a voice, may end up only partially canceled or interfered with, while others in that waveform might be reinforced. This can result in a phenomenon known as "comb filtering" (see Figure 1.13), producing an unnatural, phasing artifact in the sound.

This type of phase interference can happen for any number of reasons, but most often, it is the result of two or more open microphones being close to each other, with each of the sound sources getting into the opposing microphone.

Take the example of there being two people in a scene where we need to capture their dialogue. To do this effectively, each actor needs to be far enough away from the opposite actor's microphone so they're not picked up on that mic as well. That would be some awkward staging, positioning the actors so far apart, so the fact is that we're just going to hear both actors on both microphones. Sound doesn't have the standing to tell the director how to stage their shot, so in the end, the goal is to have each actor be strongest on their

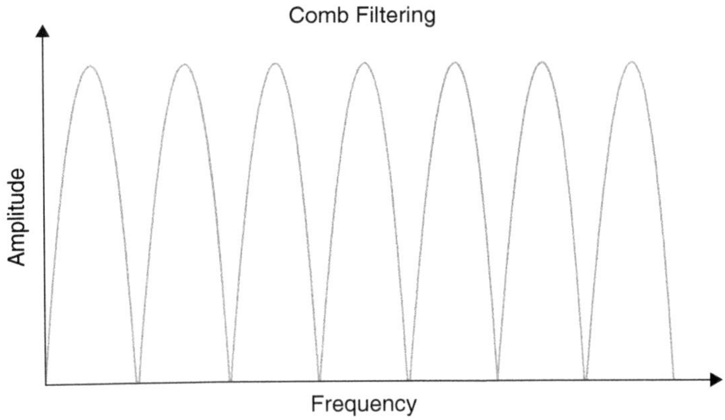

FIGURE 1.13 Comb filtering example, where a 1kHz tone is shifted 0.5 ms relative to the original, resulting in odd cancellations and reinforcements of frequencies in the upper range of the spectrum

dedicated microphone. As the two actors get closer, each becomes stronger on the opposite mic. At some point, if those two actors are mixed at approximately equal levels, we will hear the two microphones' signals begin to interfere with each other.

To try to counter this effect, we refer to the 3:1 rule, which states that the opposite microphone should be three times farther away from the subject it's NOT directed toward than it is from the one that the microphone is meant for (see Figure 1.14). Once the 3:1 rule can't be enforced because of proximity issues, it's time to figure out how to capture both actors on one microphone. (See *Media 1.4m – phasing example.*)

Dealing with and mitigating phasing issues can be a large part of production sound recording, particularly as modern sets demand greater coverage, meaning many more microphones are being deployed to capture every part of a conversation. Even in a standard documentary-style interview, it's standard practice to use two microphones on a single subject being interviewed; this is meant mainly as a safety measure but also to provide options to the post-production team.

A camera-mounted microphone isn't the best option in most cases, as the distance from the camera to the subject can vary greatly. The angle can also be less than ideal in terms of the subject being on-axis and within reach of that mic. The ideal setup is to have a boom microphone above the subject and a lavalier microphone placed directly on the person. Whatever the setup, utilizing multiple microphones ensures coverage while also creating potential phase issues – but in all scenarios, we will usually choose between the two, going with the one that sounds better and likely not having them both at equal amplitude in the final mix.

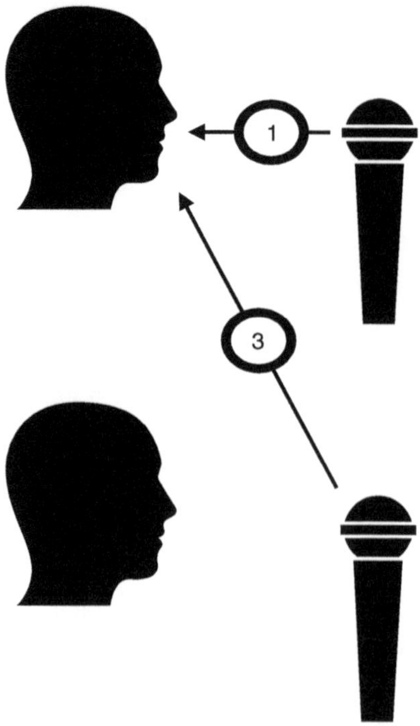

FIGURE 1.14 The 3:1 rule – microphone placement to avoid phase cancellation

On a narrative film set, shots are often designed so that it wouldn't be possible to capture all the actors on a single boom microphone or with even several booms, as the potential for those booms to interfere with the action or create shadows is problematic. Mixers often resort to "wires," which is on-set shorthand for placing individual lavalier microphones on each actor that will have lines in the scene – which inevitably means several open mics at once. This is usually fine, as lavaliers will be very close to the person they're placed upon. If we have two actors that need to be in closer proximity to each other, such as sitting close to each other or hugging as part of the scene, then the choice needs to be made to determine which microphone on which actor gets to carry the scene in those tighter moments.

1.2.2. The Mechanics of Sound – Reflections

There's simply no getting around the physics of it; sound will bounce around and be difficult to contain. Sometimes, it seems as if sound defies all our best efforts to capture and manage it, no matter how we try to isolate the elements being recorded. It can bounce off hard, reflective surfaces like glass or polished floors. When sound has a large area, such as the high ceilings in a gymnasium or a church, it can traverse the space in endlessly overlapping and uncontrolled ways.

It's easy to see why large spaces are a less-than-ideal environment for capturing clean dialogue – with so many uncontrolled reflections, it will interfere with voice intelligibility.

If, on the other hand, the location is a smaller space with a lot of carpeting, draperies, and upholstered furniture, such as a living room or bedroom, then sound reflection is going to be far less of an issue, as many of the surfaces in that space are very absorptive.

To help understand why reflections so significantly affect the clarity of sound, it helps to know that the human hearing process is based on spatial hearing – we are binaural creatures, and we hear sound in 360°, to some extent. Evolution and our survival as a species allowed us to locate a sound's source by effectively comparing the time differences between direct and indirect sound as sound reaches one ear first and then the other. Our ear and our brain work together to calculate the timing and intensity differences of both direct and indirect sound and how each arrives at each ear, providing us with the ability to localize that sound source.

To understand this, observe the life cycle of a sound wave in an enclosed space, such as a lecture hall. There are three components of that sound wave to look at: the direct sound, any early reflections, and then reflected or indirect sound (see Figure 1.15).

Direct sound is exactly as described; it's the straightest path from the sound source to the subject's ears, but there is still a lot of information for our ear–brain to calculate from that. We can very quickly localize that sound source in space, as the difference in that sound's arrival time varies from one ear to the other. It may be only microseconds, but the human brain is an efficient analyzer and can effectively locate that sound source.

After the direct sound, some early reflections are perceived as the sound wave reflects off intermediary surfaces as it travels from the sound source to the listener. For this, imagine the furniture in a space, but also the flooring, the walls, even the ceiling – anything that the sound wave contacts on its way to the listener is going to interact with the initial sound, and all those surfaces will be creating additional reflections of that sound to some degree. Lastly, there will be the indirect sound, usually consisting of reverberant reflections in the space, as the sound goes around and beyond our subject, reflecting off other surfaces in the space and then coming back to the listener. Imagine hard, reflective walls around the speaker and the listener, a whiteboard or a blackboard, and perhaps a hard, uncarpeted floor. Those additional reflections further help us localize the sound source and provide further clues to describe and identify

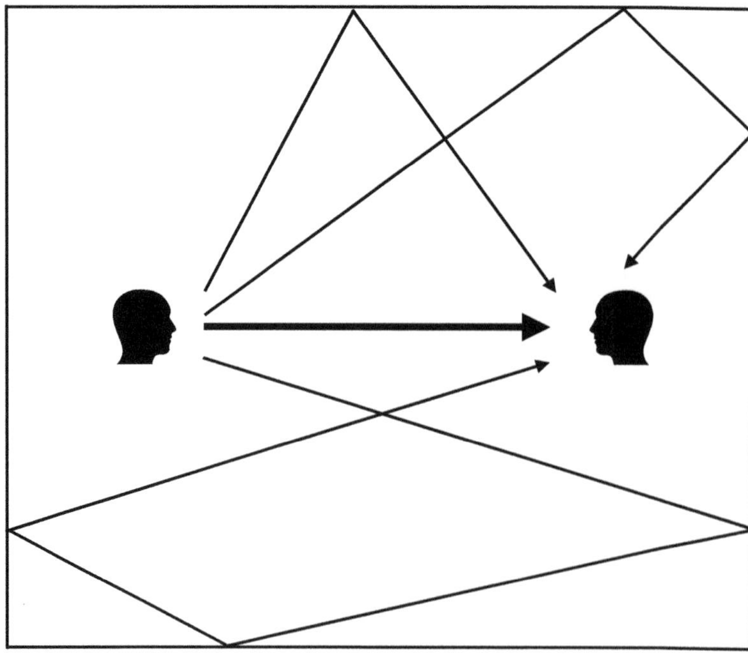

FIGURE 1.15 Direct vs. indirect sound

the space we're in. The ear/brain computational combination will quickly do that math and confirm that we are indeed in the classroom we are seeing. Even if we are deprived of sight, it would be simple to tell if we had walked into a closet, a classroom, or a gymnasium, for example, simply by the sound feedback the space gives us.

Of course, as recordists on a set, we're most concerned with capturing that direct sound while doing our best to exclude as much reflected and indirect sound as possible.

Reflections can be classified in two main ways: first, and perhaps most apparent, is what we hear as an echo, described as *discrete* or independent reflections of a sound wave. Imagine the classic scenario of shouting into a canyon, where you would hear your words coming back to you in a more or less accurate representation of the initial, direct sound. Echoes occur in large, unconfined spaces, defined as having a surface opposite the source, with no containment above and around, as a canyon would. Echoes can also occur between buildings on a city street or between other facing surfaces, where two opposite walls allow sound to reflect back to its source.

We also experience sound reflections via reverberation, where sound waves bounce off many different surfaces, resulting in a series of *diffuse* reflections, with those reflections overlapping and interfering with each other in their journey back to our ears. Reverberation occurs in enclosed, contained spaces, such as a room with four walls, a floor, and a ceiling, but it can happen in any closed-in space that keeps the sound contained within its finite volume. In this situation, there are so many reflections bouncing around and overlapping with each other in many ways that we can no longer hear any of those reflections individually; they've blended together to create the reverberation sound field we now hear instead.

Both can be useful effects to help relate to the visual environment, and in post-production, we can re-create these effects with sound processing tools to help create an environment or setting after the fact as well. To do this, however, the sound from the original production recording needs to be as clean and reflection-free as possible. This will give us far greater flexibility in post-production, as we can tailor the amount of reverberation or echo to match the visual it accompanies. Wider shots might get a heavier treatment with reverberation, for example, which can be reduced on tighter shots to further help intelligibility.

Sidebar – Sound Reflections

There are two realities when it comes to sound reflections – on set, we're trying to minimize them as much as possible. We're then adding them back in post-production, but often in new, better, and certainly more controlled ways, inflicting a new acoustic reality and environment onto the scene. It's a lot to deal with, so both professional sound stages and studio control rooms are carefully designed to give us the most acoustically accurate environment possible, with minimal reflection. Cinematic theater spaces are also designed to be as acoustically neutral as possible, allowing that detail in the sound work to shine through.

In both the theater and the control room, acousticians and architects will work to create the most stable sound environment they can. All aspects of sound are controlled, and designers of these spaces have resources in tables and charts that show the sound absorption and reflection coefficients of all the materials that might go into the design of that space.

Some of that knowledge can be applied to the production stage to help minimize those unwanted reflections from the start. Location mixers will routinely use carpeting, baffles, and other sound treatments to help mitigate sound reflections and ensure that the cleanest possible dialogue is recorded.

1.2.3. The Reality and Psychology of Sound

All these factors reveal how we humans have an extremely subjective, even biased, perception of sound. The limits on our hearing ability mean that we don't perceive all frequencies equally, with increasingly greater deficiencies in the upper and lower ends of the spectrum – but a lot of the area in between is also uneven. The simple truth is humans are imperfect listening machines, which is further complicated by the fact that individual variances from human to human are also present. This is even before addressing any individual predispositions to sound; these can be cultural, biological, or even psychological, all factoring in to make the human response to sound even less predictable.

1.2.3.1. Perception

Where humans are concerned, all frequencies are not created equal. We can see from Figure 1.16 that HHR falls off quickly below a certain frequency point, and it also decreases at the higher end of the spectrum. Low frequencies can be a factor for anyone of any age. Still, it's the higher frequencies where age, biology, environment, and general predisposition to certain frequencies and types of sound can cause increasingly varied outcomes.

Figure 1.16 also shows sensitivity at different amplitudes. We hear the frequency spectrum differently at various loudness levels, with the valleys of our

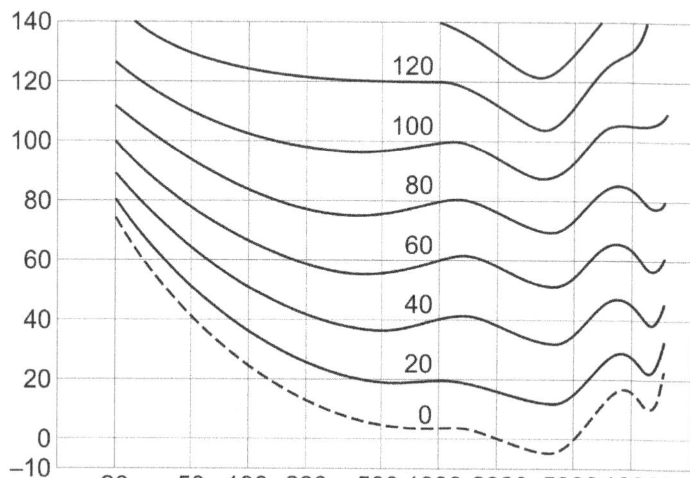

FIGURE 1.16 Equal loudness curves

Source: By Monedula – this file was derived from: Gromkost.png, Public Domain, https://commons.wikimedia.org/w/index.php?curid=90640847

perceptions getting deeper as the overall level decreases. At higher loudness levels, the trend is for those peaks and valleys to flatten out slightly. While we humans will never have a completely flat response, the differences become less pronounced at greater amplitude.

1.2.3.2. HHR – the Realities of Binaural and Selective Hearing

How we receive and perceive sound is based on our two-part ear–brain receptor system, which has proven very effective over the evolution of the human species. Our two ears, together with the brain that sits in between, can effectively receive, decipher, and then make sense of auditory inputs, all in an effective working combination that has allowed us to survive and evolve to this advanced industrialized place where human society exists.

Simply surviving was a much more challenging proposition for our human forebears, as there were several more immediate threats to our very existence than there are now. Some of our earliest basic challenges were hunger, disease, and the ability to clothe and shelter ourselves effectively. Most importantly, we had to get through our day without being eaten by something bigger and hungrier than ourselves. Our hearing capabilities evolved to help maintain a constant threat assessment, a background processing that serves to alert us to any immediate dangers. This is no mean feat for any species, especially for a species with such inadequate hearing. Our ear–brain combination helps us to identify and then accurately calculate a sound's direction and proximity, and it does this by doing some rather complicated math and perceptions along the way. As discussed, it can calculate the difference in timing between a sound's arrival at one ear versus the other and then triangulate the sound source's approximate direction and location from that data – all very quickly and accurately. The result is that we can focus our directional response to assess any possible threat, bringing in our eyes to assess further the possibility of whether that threat is dangerous and then engaging the brain further to formulate a physical response or reaction. The amygdala generates the fight-or-flight response, which decides how much adrenaline to dump into our bloodstream to aid us in our response.

Our ancestors didn't have an easy time of it against predators – we humans don't have much going for us in natural physical defenses, like the teeth and claws that many of our potential predators do, and we also don't have wings or swift feet to escape. In short, we needed all the help we could get, and even before developing tools and weapons, our hearing capability was a big part of our ability to survive those threats, serving as our first alert to sonic stimuli.

Another of our strengths lies in our ability to hear and understand each other's spoken word – this helped us develop language, allowing us to share thoughts and accumulated wisdom. That shared wisdom became critical to our survival, and together with the sharing of technology, communication was a central part of our ability to survive and, later, enjoy some evolutionary success. That discretionary hearing ability remains with us today, allowing us to hear and understand each other's spoken word in challenging situations. Binaural hearing is undoubtedly a part of this, but so is our ability to decipher the sonic stream coming at us and then quickly analyze it to decide what information to ignore and what is important to us so that we can effectively concentrate on what matters most.

For example, we've all had the experience of sitting in a relatively quiet room. In this scenario, imagine being in a kitchen and having a quiet conversation – it's late at night, so other environmental ambient noise is minimized. When we're in any environment that's relatively consistent, noise-wise, we tend to disregard whatever slight noise there may be in favor of the more important conversation at hand. Much of our attention can then be focused on hearing and understanding that communication. At the same time, our sound awareness allows us to overlook those less-important environmental noises that are also happening in the moment.

However, our awareness will be triggered if there's any change in our small sound world here, such as when the refrigerator cycles itself off. That sudden absence of sound in our sonic environment is a jarring change to our norm, causing us to notice the fridge for the first time. The background noise processing we have going on registered the shift to our conscious brain, making us note that difference and then helping us quickly assess whether that change was anything of consequence that might require action.

In a different ambient sound scenario, we're talking with a friend in a crowded, noisy bar. In this situation, we must speak loudly to be heard, but we're ostensibly there to have a good time and enjoy some good conversation despite the noisy circumstances. Humans can do this reflexively because we excel at excluding sounds that are not a part of our conversation, so our ear–brain is essentially – and efficiently – ignoring sound that isn't part of the immediate conversation. This is thanks to our selective hearing, which is a combination of processes at work.

First, we can successfully ignore some of the surrounding noise, but only to an extent, as there will still be words and phrases that won't make it through the clutter of noise. Here, our listening mode is further enhanced by context, meaning that as we follow the conversation, there's undoubtedly a subject to build from, some continuity thread informing us what the other person *might* be saying. Lastly, we humans are also pretty good at lip reading, whether we know it or not, and it's another skill that helps us further understand the conversation. Our ear–brain combo works together, incorporating the visual cues we see in the other person's mouth movements, adding further context to the words and parts of the conversation we understand. All these elements come together to help us know enough about the conversation to allow us to keep participating in it.

Keep in mind that this whole scene starts to unravel if we don't know the other person as well, which might make us less familiar with their speech patterns. The situation could be further compounded if they speak with an accent that is less familiar to our ears. We would have an even harder time understanding the conversation if we were trying to converse in a language we aren't as familiar with; we would need more accurate auditory input for help.

Ultimately, it's important to note that humans hear subjectively, whereas microphones hear objectively. So, while a microphone might be technically very accurate in how it picks up sound, we are much better at filtering and deciphering the important material from this flood of sound data. Ultimately, we perceive sound in these noisy settings far more effectively than any microphone ever could. As evidence of this, one can't simply put a microphone out in the middle of a crowded bar and expect to get even a passable recording of the voices in the conversation we want to hear. The microphone hears everything.

1.2.3.3. HHR – Strengths and Weaknesses

Further compounding the weaknesses in human hearing is our inability to hear all frequencies equally. As it happens, HHR is tailored to the frequencies where language components typically reside while also being good enough to help our prehistoric selves better evade those apex predators.

At either end of the spectrum, sounds move into areas where they are felt more than heard by humans. This is easy enough to imagine in the lower frequencies – imagine a thumping bass track or the sound of an earthquake. Low-frequency energy can sometimes seem to physically move the world around us and resonate in our core; low-frequency sound waves are felt on a visceral level. We are slightly more receptive to low frequencies (infrasonic or subsonic) but ultimately not as efficient on either end of the spectrum as in those middle ranges.

Higher frequencies hit us differently, whereas low frequencies fade away before they drop off the scale of human hearing ability. HHR is nearly universal on the low end of the frequency spectrum; it's something most hearing humans have in common. On the other end of the spectrum, high frequencies start disappearing the higher we go, a phenomenon further compounded by whatever condition our hearing is in, before dropping away entirely as they go higher and, eventually, completely beyond our effective hearing response range. As they ascend, those frequencies might be experienced more as pressure on our ears before they finally go beyond our hearing range. Unfortunately, many of our predators were better equipped to hear those ultrasonic frequencies than us.

In the end, we humans have a lot working for us in terms of hearing perception, but there's also a lot working against us. As sound designers, we need to be aware of the inefficiencies in the human hearing process while also being able to account for and accommodate them.

1.3. GEAR – AN OVERVIEW

Technology marches on, but the main concepts remain constant. A brand-new microphone might be functionally the same and utilize the same basic technology as a microphone from the 1940s, for example, and a speaker is still functionally a speaker. That said, those technologies continue to advance and be refined over time, with the technological means of executing many of the steps between sound generation and reproduction evolving and changing as well, sometimes in significant ways. However, the fact remains that, even with so many refinements and technological advancements, someone from the Sound Department of a film in the 1950s would be able to understand the functionality and workflow of a Sound Department in the 2020s, as so many of the concepts have remained the same.

1.3.1. Microphones and Transducer Types

Technology has allowed us to develop incredibly flexible, purpose-built microphones in the modern age. There are microphones for drums, voice, and other instrumental mics, as well as microphones for even more specialized tasks. Many refinements of design, application specialties, and models are available to fill a microphone locker with a plethora of refined and specialized microphones. But the basic concepts have remained the same at the heart of it all.

At its most basic, a microphone is built to acquire acoustic energy and convert it into an electronic signal that can then be stored, amplified, or otherwise manipulated. While many different designs and technologies are currently in circulation, and the design refinements continue, they all perform the same function in transducing one form of energy (acoustic) into another (electronic).

In Figure 1.17, we see a simplified diagram of a typical dynamic microphone design. Acoustic energy – that is, sound waves – impacts the diaphragm of the microphone. The diaphragm element is typically a very thin membrane, perhaps a small, circular disc of metal or plastic, similar in design concept to a drumhead. As that sensitive element reacts to and is dynamically reshaped and deformed by the subtle changes in sound pressure inflicted upon it, it responds and moves. In this diagram, the diaphragm is attached to a coil of thin wire, which will then move slightly between the poles of a magnet – a typical design concept. The diaphragm's motion, in turn, will induce fluctuating changes in

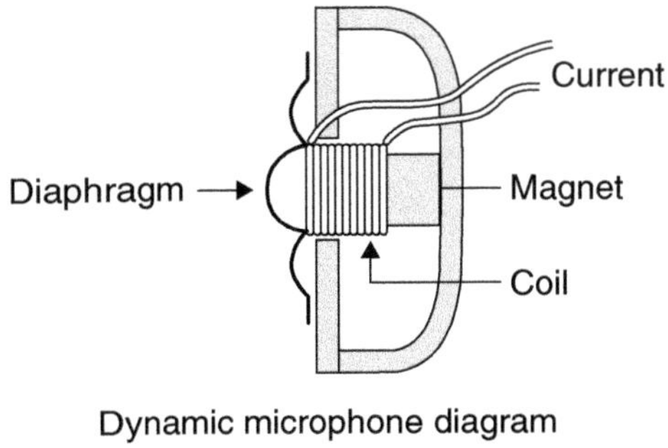

FIGURE 1.17 Basic dynamic microphone design

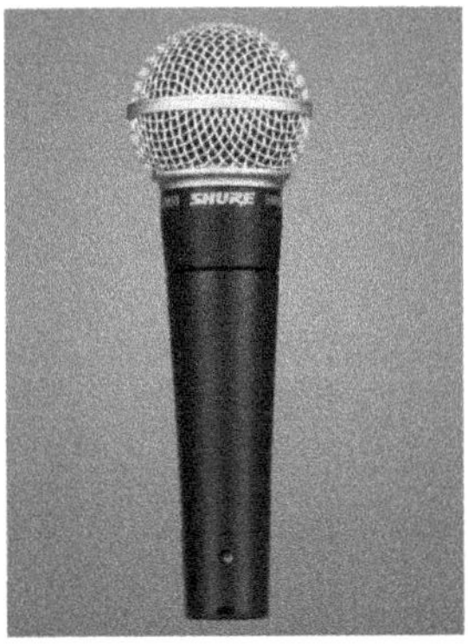

FIGURE 1.18 The venerable Shure SM 58 microphone, a popular example of a dynamic microphone

Source: Permission Shure Incorporated.

electrical current through a conversion process called transduction. Those linear changes in the electrical field can then be stored, amplified, or otherwise manipulated.

Figure 1.17 shows a very basic *dynamic* microphone design concept. Dynamic microphones (see also Figure 1.18) are typically simple in design and robust, allowing them to withstand extremes of climate, rough handling, and, most significantly, high sound pressure levels. They don't require external power to be heard; they have a high enough output that their signal can be utilized directly from the microphone.

Condenser – or, more appropriately, capacitor – microphones are designed to be more sensitive (see Figure 1.19). These microphones require an external power source and an electrostatic transduction design, with two charged metal plates forming a capacitor that maintains an electrostatic charge. One plate is fixed, while the other is designed to move in response to incoming acoustical

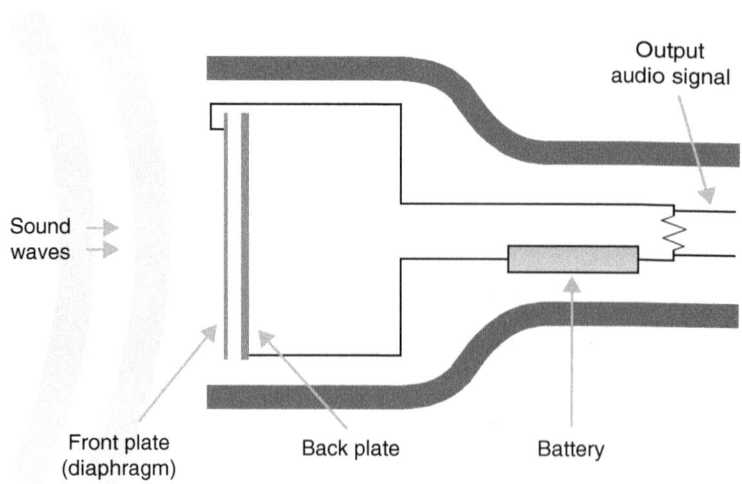

FIGURE 1.19 Basic condenser microphone design

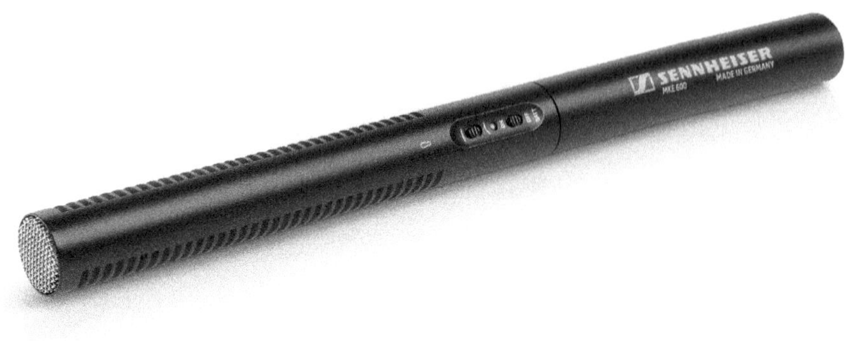

FIGURE 1.20 Sennheiser MKE 600 microphone

Source: Photo courtesy Sennheiser.

energy, working as this microphone's version of a diaphragm. As spatial variations occur between the two plates in the element, the resultant modulation in the electrostatic signal generates a sound signal.

Their output is much lower – those voltage changes can be minuscule, particularly when compared to a dynamic microphone's output, so they do require some amplification to get the output amplified to a usable level. These microphones are more sensitive to changes in sound pressure, and more accurate over greater distances than a dynamic microphone element. As a result, they also have a far more accurate frequency response – two factors that give these microphones a greater and more accurate "reach" than most dynamic microphones.

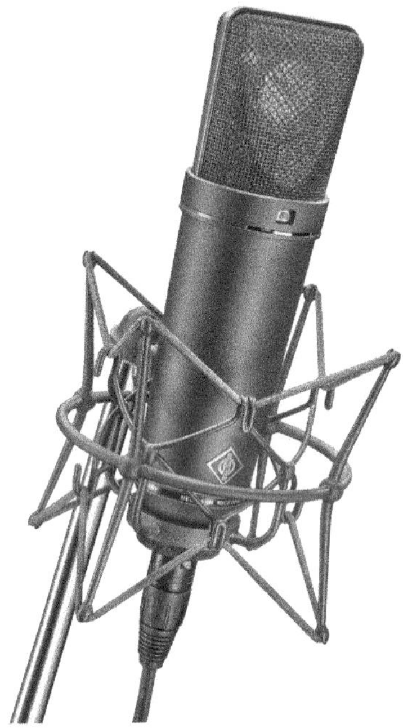

FIGURE 1.21 Neumann U87ai microphone

Source: Photo courtesy Georg Neumann GmbH.

The Sennheiser MKE 600 (Figure 1.20) is an example of a microphone with greater reach – it can be positioned at a greater distance from a subject while maintaining an accurate frequency response. The Neumann U87ai (Figure 1.21) is a high-quality studio mic. Its larger diaphragm contributes to its accurate frequency response, making it an iconic microphone for both voice and instrumental recordings.

That sensitivity makes a condenser microphone ideal for use on a set, where we will likely be farther from the voice source than would otherwise be ideal. An exception to that would be in the case of lavaliere microphones, like the DPA 4660 (Figure 1.22). While they will be closer to the subject being recorded – placed directly on the person – accuracy of frequency response is of the greatest importance to get the best possible representation of the voice.

There are other microphone design types, but many are less desirable for set use. Ribbon element microphones are one example – while a much older design, they also have excellent fidelity and extremely low self-noise. The downside is that they are also extremely delicate and simply not durable or dependable enough for the rigors of set work.

Microphones come with a broad variety of pickup patterns, also referred to as polar patterns. As with everything else, there is a pickup pattern to suit

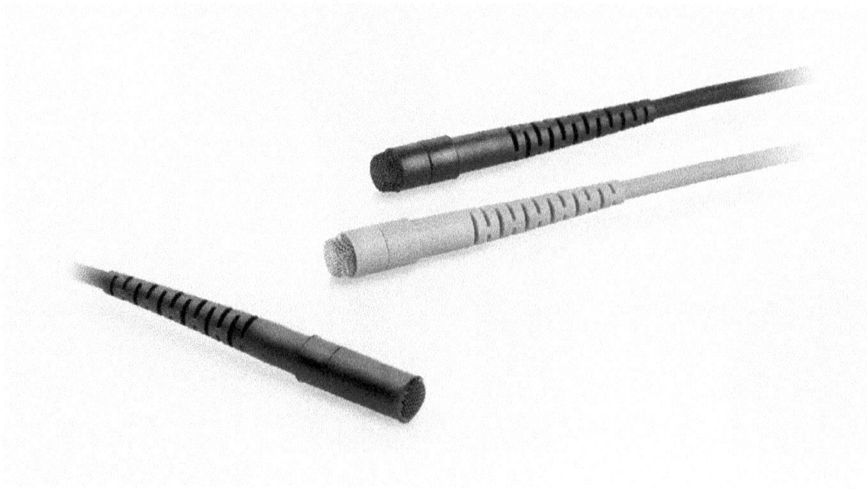

FIGURE 1.22 DPA 4660 lavaliere microphone

Source: Courtesy DPA Microphones, Inc.

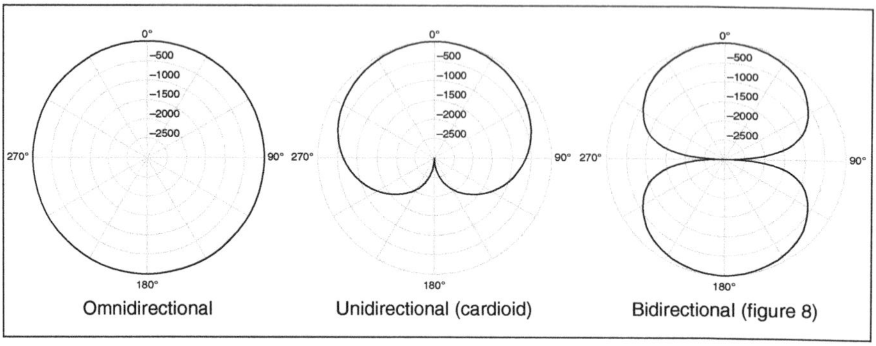

FIGURE 1.23 The three basic microphone pickup patterns

any application, but they fall into one of three main categories: omnidirectional, unidirectional, and bidirectional patterns (see Figure 1.23).

An omnidirectional microphone has a 360° sphere of sensitivity around the microphone, meaning you don't have to point it directly at the sound source; it "hears" equally in all directions simultaneously, with no axis point or central point of sensitivity. This pickup pattern doesn't lend itself to isolating the sound source well, as it will capture all sound events within its sphere of coverage.

The unidirectional pattern is the pattern to choose when we want to record a sound source while excluding as much of the environment around it as possible. The degree of directionality can vary, where some can record in a wider version of the pattern – ideal for voiceover recording and environments that are more controlled, like the recording studio – or they can go to a much narrower pickup pattern, in what is often referred to as a "shotgun" microphone. Most microphones at the end of a boom pole on set have a narrower field of pickup, designed to be aimed into a scene and capture the dialogue as effectively as possible while excluding all else.

Ultimately, mixers on location will have an arsenal of microphones with a range of pickup patterns ready for their boom operator to use, depending on the situation and the environment. A busy city street demands a narrower pickup pattern in the ongoing effort to exclude as much of that noisy environment as possible, whereas a broader pattern would be better used in more controlled settings, like interior scenes or a soundstage.

The bidirectional pattern doesn't have any real application on set on its own – having two fields of sensitivity makes it difficult to apply to most film set scenarios – but it may have some applications as part of a multi-microphone array. A bidirectional microphone picks up on two faces that are 180° opposed from each other, with maximum rejection at the nodes that are 90° off from those. Think of it as a talk show microphone, where two people can face each other with the microphone in between, able to pick them both up equally. It can also be used as part of a stereo or quad setup, where its signal is combined with other microphone patterns.

Lavaliere microphones are frequently used on a film set, where they are often applied in an omnidirectional pattern, which wouldn't seem the obvious choice for that environment. Ostensibly, we want to exclude as much of the noisy world around us as possible, and an omnidirectional microphone would seem counterintuitive, at best – but there are other factors at work here.

In a typical film set application, there is a lot to help isolate a lavaliere from the noises of the world – clothing and other wardrobe, and even the body it's attached to all work to limit the practical sound field or "reach" of the microphone. There's more to choosing the omnidirectional pattern than that, however.

Proximity effect is a phenomenon in directional microphones where low frequencies become accentuated, even exaggerated, as the sound source gets closer to the microphone. Some radio personalities and announcers embrace this and make it work in their favor by intentionally getting closer to the microphone, using that boost to emphasize the bass in their voices and make them sound bigger than life. However, at other times, it can be detrimental when the sound source gets too close, as it may remove perspective and any sense of environment from the sound signal. That DJ voice is certainly rich in bass frequencies, but it lacks perspective and any sense of space; it isn't a realistic way to represent the spoken word in conversation. Being so close to the microphone also makes it prone to wind interference and the dreaded "p-pop," where the microphone picks up puffs of air, turning them into low-frequency "hits" on the mic.

However, not all pickup patterns have the same proximity effect issues, with cardioid microphones being the most prone to it. Omnidirectional microphones are essentially immune to the proximity effect, so utilizing this pattern in lavalier microphones makes them the ideal choice for placement in some challenging, potentially bass-heavy situations. On top of all that, the constraints of working around wardrobe and other issues on set leave limited options for placement, forcing us into areas that, even though they may keep the mic out of sight, may be far less than ideal from a sonic standpoint. Lavaliere microphones are ideal since we can't always point them in the right direction when constrained by issues of wardrobe and visibility on camera. That omnidirectional pattern gives us the most flexibility in the end.

Lavaliers can certainly be unidirectional as well, and although those patterns may also have some application on a set, directional lavalieres see more application on open sets, such as broadcast news or talk shows, and in live sound reinforcement situations.

1.3.2. Production Recorders

Once a suitable sound receptor has been selected, we need a place to store the information it's bringing us, so we move from gatherers (microphones) to the storers. There are many ways to store an audio signal, ranging from the old days of wax cylinders, vinyl records, and magnetic tape to the current technology of disc- or flash media–based recorders.

Flash media recorders lend themselves well to sound gathering in the field. They are of a different design than what we might find in the studio, as these recorders need to be compact, self-contained, robust, and able to perform under a variety of sometimes less-than-hospitable conditions. Some actual tape-based machines are still in use, but analog tape has become scarce, and most location recordists have come over to the digital recorder side. With virtually no internal moving or mechanical parts and access to cheaper and far greater media storage capability, this technology has many benefits, with the latest generation proving the durability of the platform.

The on-set tape-based recorders were the norm for nearly 60 years. Highly refined and durable (but also heavy), they certainly had their limitations, the biggest being their limited number of tracks. These decks used ¼" open reel tape, capable of recording only one or two audio tracks. They were mechanically driven and had a much smaller media storage capability than flash media provides – a reel could only hold 16 minutes of tape on a 5-inch reel or up to 32 minutes on a 7-inch reel. A watchful eye was needed on set, and among a mixer's many job skills was the ability to quickly change reels when the tape level got down to a certain point. Fortunately for the Sound Department, camera film

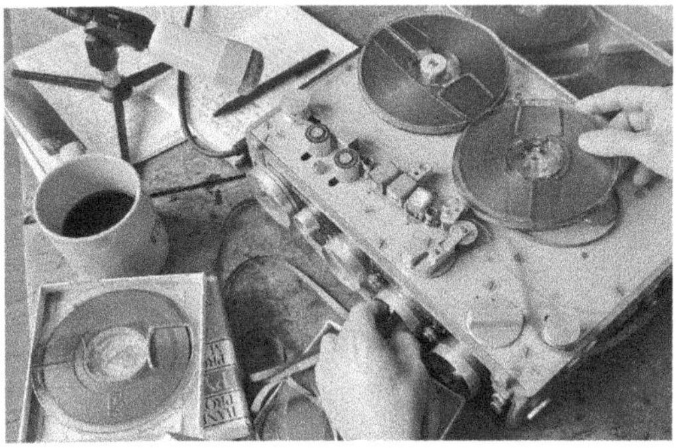

FIGURE 1.24 Nagra reel-to-reel field recorder

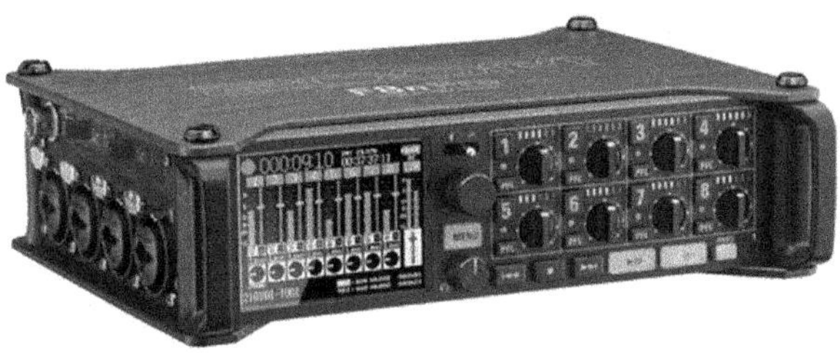

FIGURE 1.25 Zoom F8n Pro field recorder

Source: Permission Zoom, Inc.

loads were even shorter in duration, so there would be plenty of opportunity to make that reel change happen.

Despite the bulk, they were robust machines, and there are many stories about how the legendary Nagra (see Figure 1.24) persevered through the worst and most adverse conditions.

Currently, many audio recording decks, like the Zoom F8n Pro (Figure 1.25) and the Zaxcom DEVA 24 (Figure 1.26), use flash media, giving them the capability to hold several hours of audio information on a piece of media as small as a postage stamp. They're also capable of multitrack formats, going well beyond the one or two tracks on the reel-to-reel format of the Nagra. As digital technology advances and becomes even more ubiquitous, it also becomes less expensive to manufacture. No matter the price point, there's likely a field recorder to fit most production budgets.

Another essential tool in managing all those tracks is taking control of all the microphones by directing their inputs through a mixer interface, like the Zaxcom Mix-16 (Figure 1.27). This allows the mixer to send microphones to the

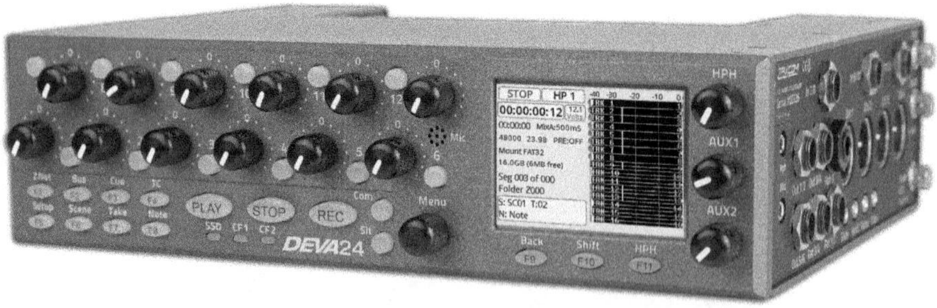

FIGURE 1.26 Zaxcom DEVA 24 field recorder

Source: Photo courtesy Zaxcom.

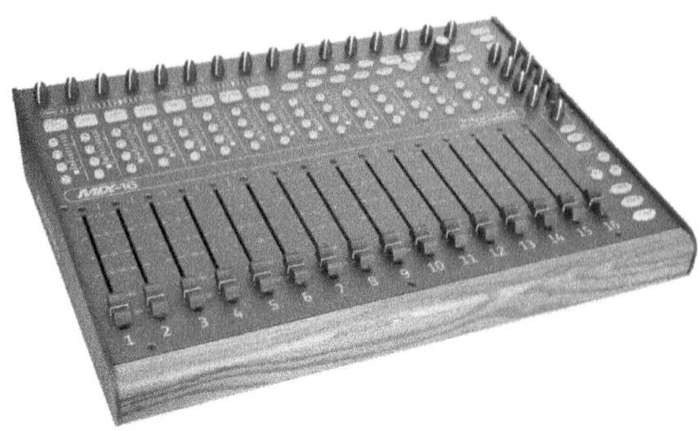

FIGURE 1.27 Zaxcom Mix-16 control surface

Source: Permission Zaxcom, Inc.

appropriate tracks while simultaneously having the ability to create side mixes – also called sub-mixes – simultaneously.

The mixer on set is primarily responsible for acquiring clean dialogue, but they're also serving several departments at the same time. The post-production team needs all dialogue tracks to be isolated from each other. At the same time, the director needs a coherent mix that is representative of the on-camera dialogue provided with the rushes of the previous day's work – they simply don't have time to work through the many tracks captured in a day.

Therefore, the second most important task for the mixer is to provide a coherent, working mixdown of all those tracks.

The picture edit team will need a concise mix to do their work, while in audio post-production, the dialogue team will need all the isolated tracks to do their job effectively. Having all those microphones isolated on individual tracks is a requirement for the sound team, but it's problematic for the picture editor to try to make sense of them all. The editing team simply needs the best representation of the dialogue track, which helps streamline the offline editorial process. There's plenty of time to revisit those isolated tracks later, if necessary, and that work is best left to the Sound Department.

It's also a source of pride for production sound mixers for their mix to be as close as possible to what is ultimately needed in post. Being familiar with the script, the choreography of the action on set, and how the set environment sounds, they have the best perspective of how all that production audio *should* sound. A properly executed production mix will satisfy most of our audio post-production needs as well – it's only when we need to reinforce or replace a given line that we might need to delve into those iso tracks further.

This is a level of coverage and accuracy that didn't exist in the past, becoming standard practice only as multitrack and multi-microphone setups have become the norm. At the same time, these more complicated setups require better resource management and organization. Having so many microphones means they will need to be routed efficiently and with a flexible system that can be reconfigured to the needs of any given production day.

Recording dialogue between two people is easy enough, in concept – standard practice would mean placing wireless lavalieres on each actor, with a single boom above to capture both actors and the acoustic signature of the space they're in. But even with such a simple setup, those three tracks require three separate inputs to be "tracked," or recorded, separately. On set, the production sound mixer will make their best assessment between the three mics, balancing them into a concise mix for ease of use in the editorial processes to come.

Sidebar – Sample Rate and Bit Depth

There are a few choices to make when setting up that recorder and heading out to record audio, and one that often needs clarifying to new mixers is the question of sample rate and bit depth. There are two sides to this digital coin, the first being the sample rate.

The sample rate refers to the gathering of analog audio and converting it into digital data, with the number of digital samples converted per second being the sample rate. Usually, professional audio in film and video is sampled at a rate of 48,000 samples per second. Depending on the options available on the recorder, there may be other sample rates to choose from. Another common sample rate is 44,100, notated as 44.1kHz, a sample rate that is the standard for CD audio. Some systems and recorders have higher rates available, appearing in multiples of the

44.1kHz or 48kHz rates. It's not exactly uncommon to see sample rates of 88.2kHz and 96kHz, with studio-based systems going even higher, to 176.4 and 192kHz sample rates.

In our world of "more is always better," that technically checks out. Still, it's important to note that for one to hear a difference between 48 and 96kHz, only a truly stellar, state-of-the-art sound system would reveal that – and even then, any differences would be subtle, at best. True audiophiles will insist they can hear that difference, especially at the highest rates, but distribution paths to the public remain elusive at this time. Further, that expanded sample rate also means the media fills up faster, so there is a (nominal) cost associated with that increase in the data stream.

The gains to be realized in different bit rates are more readily apparent. Essentially, the sample rate corresponds to the linear axis of time, with bit depth representing amplitude in the vertical. But just as decibels are a measurement system complicated by its ratios, bit depth can also be a challenging concept to grasp.

Most recorders have an option to switch between 16-bit or 24-bit audio. At a 16-bit rate, each sample has 65,536 different amplitude values available, which would seem to be a lot. However, at 24 bits, there are 16,777,216 amplitude values per sample at your beck and call. At 16 bits, that would mean a dynamic range of 96 dB, measured from the noise floor to 0 dB Full Scale, or FS. At 24 bits, the dynamic range increases to 144 dB. In the digital realm, where noise isn't usually a factor, either bit rate will be sufficient, but the key is in the available dynamic range: 96 dB (at 16 bits) seems like a wide swath of amplitude values to have readily available; it's close to our effective hearing range as humans. At 144 dB, 24-bit audio well exceeds human hearing range if 130 dB is our threshold of pain.

The limitations of 16-bit only become apparent when quieter audio signals need to be brought up in level – if recorded at a low enough level, any increase will reveal the noise floor, presenting in an artifact known as dither. Dither occurs when the bit resolution decreases at lower amplitudes, with a modulating noise floor becoming apparent as the available number of bits diminishes. That noise floor is the steps in bit depth resolution being revealed – similar to how we might begin to see the pixels in a picture that has been extremely magnified. The loss of resolution of an audio signal that we see in Figure 1.28 results in a grainy, noisy quality that modulates with the audio signal.

What's great about 24-bit recording is the increase in available amplitude, making any loss of bit depth less likely to happen. Low-level

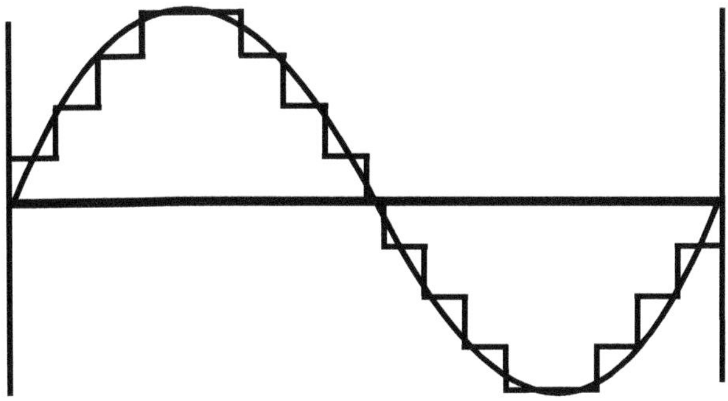

FIGURE 1.28 Loss of bit depth at low resolutions

signals can be easily amplified to usable levels without increasing the noise floor. Very simply, it exceeds human hearing capabilities, going beyond our threshold of pain at the top and well below our threshold of hearing on the lower side.

In this context, it's important to note that the North American and European broadcast audio standard requires a sample rate of 48kHz and a bit depth of 16 – more than adequate for a high-fidelity listening experience. Film bit depth often utilizes 24 bits, which provides a greater dynamic range. Streaming outlets can be a bit of an outlier, where there is a lack of standardization. Depending on the platform, sample rates can be either 44.1 or 48kHz, with bit depths typically settling in at 16 bits. There are exceptions to the rule – different factions in the streaming world have their own best practices. In all cases, it's important to know the deliverable requirement before embarking on the project. When in doubt, we work at the higher rate of 48kHz and 24 bit – it's always easier to convert downward, when needed, without the technical compromises that may result from trying to convert a lower resolution upward.

Increasingly, workstations and field recorders utilize 32-bit (float) processing. This results in a massive increase in dynamic range to 1,528 dB, with 4,294,967,296 values available. These systems don't operate at a fixed rate of 32 bits, however, but have the option to tap into those additional data as needed – hence the term "float," with an additional 8 bits accessible beyond the 24 bits normally available. The question many have is why – the resultant 1,528 dB of dynamic range is so far beyond the capabilities of human hearing as to seem ludicrous. The upper limit is 770 dB above 0 dBFS on a digital meter, while the lower limit is 758 dB below that.

Quite simply, it gives the engineer room to operate, with a wider margin for error. Consequently, audio input levels can be pushed well beyond what a 24-bit system would generally be capable of, and extremely low-level signals can be boosted seemingly forever without any noticeable increase in noise. Additionally, overloading while recording is virtually impossible – when the signal exceeds the preset level expected, the system can simply make additional bits available to absorb that sudden increase without the threat of distortion.

1.3.3. Digital Audio Workstations

The mixing console, or mixer, is at the heart of any recording studio or mixing stage. In the pre-digital era, it was the central connection for all audio going through the room while representing the most significant capital outlay on a piece of equipment in the space. The mix sources – either magnetic film dubbers or, later, multitrack tape machines – would be routed through that mixing surface, which could also tie in any number of signal processors while also sending audio to any number of destinations, including final output to the format ultimately needed for the soundtrack mix. The output was also routed to the monitors – the speakers being used in the space – with the option to switch between both speakers and sources.

It is in this space where technology has evolved the most and continues to develop at the fastest pace. Along the way, the modern Digital Audio Workstation – or DAW – has replaced the entire complicated, expensive analog setup while effectively democratizing the audio post-production process by making it more affordable, flexible, and accessible. Previously, a very large financial investment would have been tied up in technology and related hardware. By contrast, the modern audio workstation is essentially a virtual mix stage contained within some software, available to anyone with even a modest PC.

All those tracks that would have previously resided on tape or magnetic film are now contained within one system, with the mixing interface integrated into the multitrack recorder and a virtual workspace for all the editorial and routing tasks necessary.

As workstation technology and capabilities evolve, the number of manufacturers and the diversity of products are growing. Many post-production audio workflow concepts translate between workstations, just as they all emulate (to some extent) the analog terminology and workflow from that largely bygone era. The choice of workstation software often comes down to personal

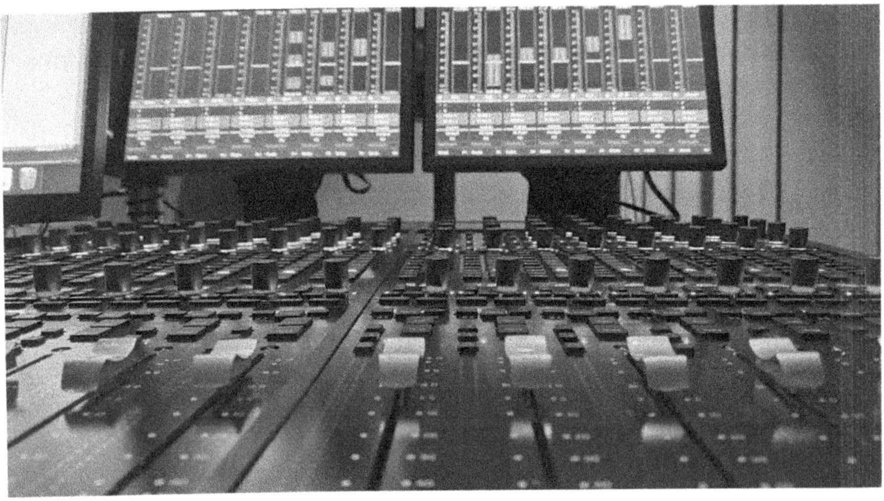

FIGURE 1.29 Pro Tools with S6 console

Source: WGBH Educational Foundation.

and technical preferences around the operating system, personal ergonomic preferences, features, overall facility of the software, and the needs of the work being done. For example, some systems are more music-centric, whereas others are better suited for work with visual media. The digital workstation concept is inherently scalable and adaptable to the task at hand, like the Pro Tools S6 console we see in Figure 1.29.

Workstations for sound post-production for film or video generally consist of a central recording and media storage system, facilitating interaction with the media through various interfaces and user options. At its core, a workstation provides the operator the essential ability to edit and otherwise manipulate audio media tracks while also giving them the means to adjust the relative levels of that audio media. As software systems become more involved, the integration of more elaborate signal processing and synchronization with video comes into play, as well as signal input, output, and routing options. This can be as simple as a laptop running sound editing software or as complicated as a purpose-built CPU tied to a multifunction console with networked media storage for large-scale projects.

For other processes outside of audio editing, an area of the program will provide access to mix functions and signal processing through various plug-in processors and another space to facilitate the routing of the audio signal. One can access the input and output signal path from the system but also manage signal flow within the system and the project itself.

At this point, things can get quite complex; the routing matrix provides access to the audio signal path in many ways, typically through the management of individual tracks and fader groups.

The main sound categories of dialogue, music, and effects will further dictate signal output paths. These can be broken out more granularly within those sound macros, depending on the specific needs of the project. Taken as a whole, this collection of audio tracks and signal processing constitutes what is commonly known as the track build.

While the mix functions are represented within the digital audio program, it's only through the hardware interface that we acquire access to those functions in a hands-on way. Mouse-driven faders on a screen simply aren't as intuitive as grabbing a double handful of physical faders on a piece of hardware, bringing those functions out to a physical work surface. Several manufacturers have accommodated this by developing digital console interfaces that emulate the earlier analog consoles in terms of look and feel. In this digital domain, they're just big, complicated control surfaces – audio doesn't travel through them as it did in the analog days. The modern console, with all its faders and knobs and displays, is a software controller that converts user input into data, allowing the user to access and manipulate audio within the system. Essentially, the modern digital console is a large and very complicated, multifunctional computer mouse.

1.3.4. Signal Processors and Plug-ins

A key part of wrangling sound media to do our bidding goes beyond providing a place to store audio track media; it's in the ability to manipulate and otherwise bend that media to the project's needs. Many tools are available in the signal processor toolbox, and they can be used to enhance, correct, optimize, and alter sounds via any number of different processes. These are often all lumped together as simply "signal processing," but they come in categories that define what aspects of the sound signal they affect. It could be any one of several sound processors performing any number of different processes on the audio signal, often simultaneously and in conjunction with each other. Fortunately for us, these tools are no longer hardware-based but available in the digital domain, where they're more easily accessed – they're just so many lines of code written to process sound.

In the analog domain, signal processors are physical pieces of equipment consisting of purpose-built circuits and processors, packaged in a (usually rack-mountable) chassis physically located in the control room and wired into the audio console (see Figure 1.30). Although this connection mode is somewhat antiquated, many of those devices remain in use today – some are still very highly sought after for their unique sound qualities. The fact remains that

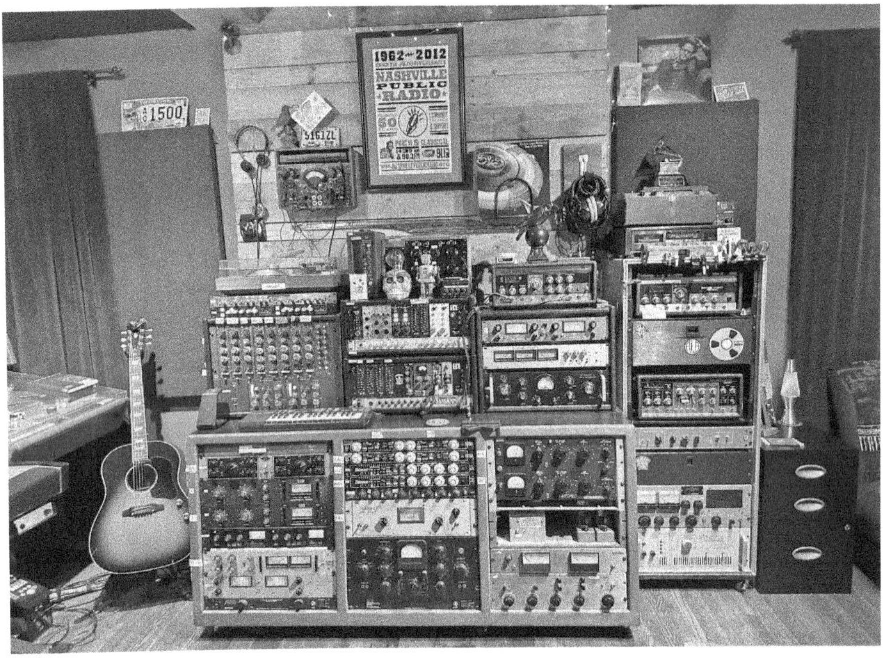

FIGURE 1.30 Analog signal processors in the equipment rack of Sputnik Sound in Nashville, Tennessee

Source: Photo courtesy Mitch Dane.

they occupy a lot of space, consume a lot of power, can be very expensive to acquire and maintain, aren't very portable, and can be very cumbersome to access and deal with when compared to their digital counterparts.

There is some merit to the sound quality issue, as many of these devices have a unique sound signature or fidelity that can be apparent to even the casual listener. However, the fact remains that a well-conceived plug-in can often approximate the same process without taking up any physical space in the studio environment and at far less cost. The arguments for or against physically present, hardware-based processors can become esoteric and are often overblown, much like the whole vinyl-vs-digital argument around LPs. Not to take sides and proclaim either to be right or wrong – the bottom line is there's no good argument against any user's preference or comfort level; it's all about perception and personal preference.

Regardless of the mode of processor – physical device vs. plug-in – audio signal processors generally fall into one of four categories, depending on the work they do.

Spectrum processors are ubiquitous and often the first stop in a typical signal processing chain; they affect and process in the frequency domain. Also called equalizers – or EQ for short – they are used to adjust the spectral content of a sound to fix deficiencies or enhance strengths within the signal, accomplished by first isolating, then boosting or attenuating, specific frequencies within the spectrum.

One basic equalizer type is a filter, a processor designed to remove or attenuate specific frequencies. Such signal processors operate in a cut-only mode, working to reduce offending or intrusive frequencies, such as removing a low-frequency rumble that might be present on location. The filter designation then implies an attenuation-only processor. In contrast, a full-function equalizer (as in Figure 1.31) can boost as well as attenuate frequencies – a far more effective means of equalizing a sound signal. A typical EQ isolates the signal into multiple frequency bands, allowing it to affect more than one area of the frequency spectrum at a time.

Our next category is *amplitude processors*, and it shouldn't be hard to guess what they do – they process and affect the signal loudness domain. By that definition, any fader or gain controller could also be considered an amplitude processor, but in this case, the discussion is around more advanced and automated processes within a plug-in or device.

One of the main tasks that post-production mix personnel have is taming and better controlling the individual dynamics of all the tracks set out before us, making relative adjustments so those tracks mix and work together better. Dialogue can literally go from a whisper to a scream within the same scene,

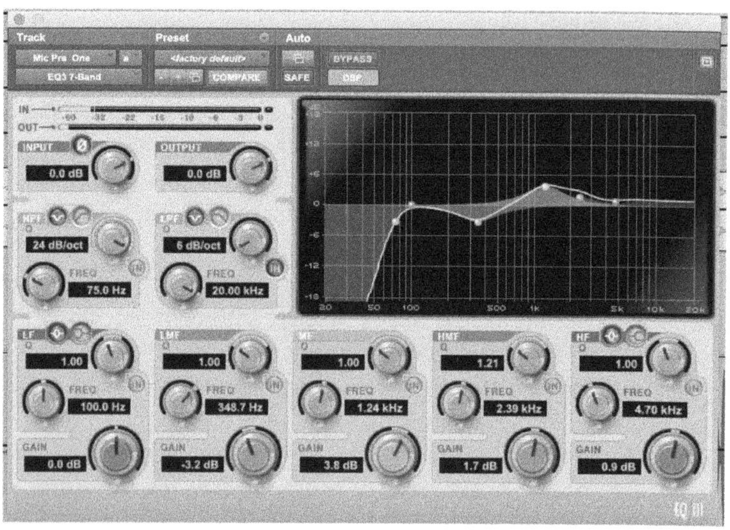

FIGURE 1.31 The equalization plug-in that is part of the Pro Tools software package

Source: AVID Corporation.

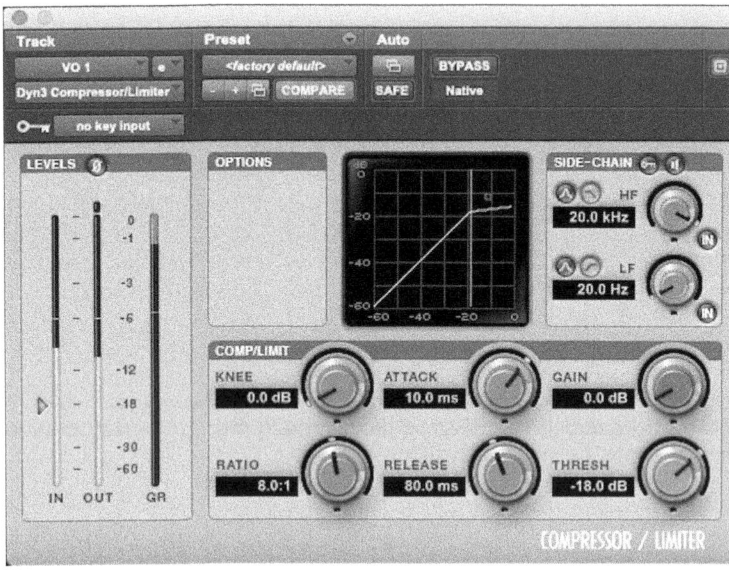

FIGURE 1.32 The dynamics processor plug-in that is part of the Pro Tools software package. Like the equalization plug-in, all parameters are adjustable and automatable

Source: AVID Corporation.

and while there's always a fader to manage that loudness discrepancy and get it under control, efforts to do so – seamlessly and transparently – are always going to be challenging.

Harnessing that signal with the help of a dynamic compressor will help even out the range within the signal. By setting a threshold at a point slightly above the average level of that signal (see Figure 1.32), a compressor can effectively lower those loud peaks as they occur, reacting far more quickly than any operator could with a fader. When set up properly, this process is mostly transparent but will also have potential consequences. As the compressor works to lower the peaks, it can make the valleys – those quieter, more whispered lines – seem louder in comparison (see Figure 1.33). This happens because the dynamic range has been effectively reduced or compressed. Having multiple individual compressors on the dialogue tracks will allow control of all of them at once and help manage them with broader strokes. Through this, the track can be managed into a more uniform level, evening out those peaks and valleys while freeing the mixer to concentrate on the entire mix in a broader sense.

Dynamic processors aren't restricted to simple compression processes; they also have other functions. There are also limiters, which are essentially more aggressive compressors. Whereas a compressor can bring those peaks down – a little or a lot, depending on the settings – a limiter is

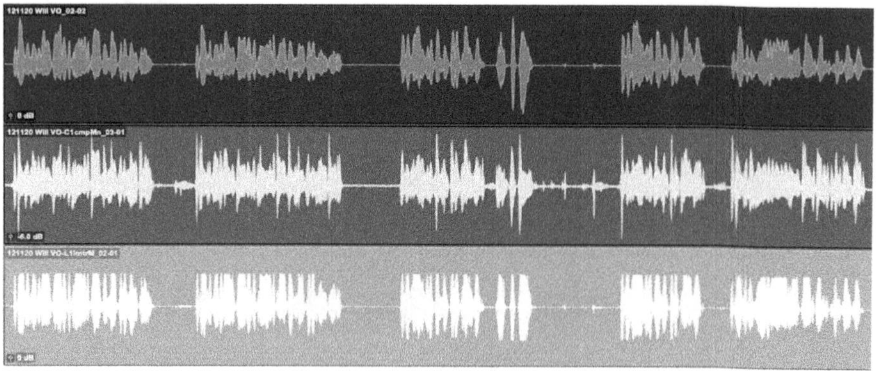

FIGURE 1.33 Compression examples – the first timeline is unprocessed voice, the second timeline shows strong compression processing in place. The last timeline shows heavy limiting – note the flattened peaks

the audio equivalent of a hard ceiling, where sounds can only go as loud as the threshold setting but can't surpass it (see Figure 1.33). This is certainly helpful in some situations and can serve as a backstop to prevent the overloading of a track or the recording of a signal. When the signal gets too loud or *hot*, as we say, that limiter kicks in and keeps that signal below the point of distortion.

Dynamic processors can also work in the opposite direction – those types of processors fall into the category of expanders and noise gates. These plug-ins will further *attenuate* audio that falls *below* a set threshold – a process that can be helpful in certain situations, such as noisier environments. The expander can be set to reduce ambient room tone or background noise during gaps and pauses in the dialogue, which can help make a noisy location seem slightly less noisy. It's important to note that this is a passive process, and even the best expander isn't selective enough to know how to keep just the dialogue material while attenuating the rest. If dialogue happens to dip into the noise floor, and noise and dialogue are at equal levels, the dialogue will be attenuated along with everything else.

Like the differences between a compressor and a limiter, there is a similar difference between an expander and a noise gate. An expander provides a nuanced level of control and will allow some of that noise to remain, ideally making the attenuations seamless and less noticeable. A noise gate, by comparison, is like an off switch – closing down any signal below the threshold setting and not allowing it to pass through the processor at all. This can sometimes sound completely unnatural; as one can imagine, it distracts the ear when all the sound modulates with voice and then disappears between sentences, words, and sometimes even syllables. Noise gates and expanders should both be used with care.

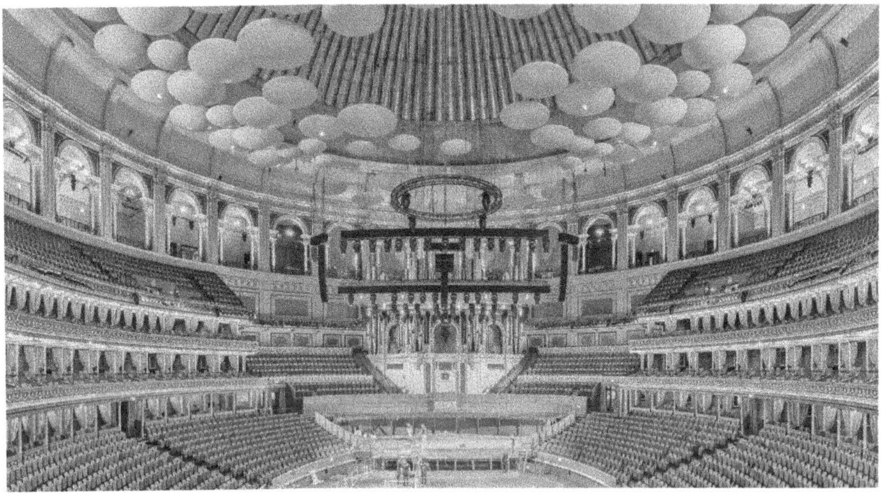

FIGURE 1.34 The Royal Albert Hall is a large space, renowned for its sonic character

Our next category of signal processing is *time processors*. Time processing is an inclusive term that includes a range of plug-ins affecting the time domain, adding a sense of space and time to the original signal. Some time-based processors affect the reverberation and echo envelope, while others can alter pitch or add modulation effects such as phasing and flanging. Putting theory into practice, the reflections and echoes discussed earlier in the chapter can be recreated with a reverberation time processor. This allows the mixer to effectively model any enclosed space after the fact but with an additional degree of control that wouldn't exist in the physical space. One example would be a classic concert hall like the Royal Albert Hall in London (Figure 1.34). This hall has a classic acoustic signature that makes it an excellent venue for all sorts of music, but it represents an extremely limited resource – it simply wouldn't be practical for a mixer to book time in there just to get a good dose of its reverberant qualities.

Fortunately, a variety of plug-ins that can successfully emulate that space (and many others) are readily available for a very modest cost and with the bonus of a lot of additional flexibility in features. It's possible to create and preserve the basic qualities of the hall, but they can also be modified; the reverberation time can be extended or shortened, and other aspects of its acoustic signature can be altered. For example, by introducing delay and other reflection elements, the signature reverberation of the hall can be taken into another realm, expanding on the foundations of an already great-sounding hall. Even nonrealistic echoes and pitch-changing elements can be introduced, ultimately moving away from the actual acoustic properties of the Royal Albert Hall and creating something truly unique in the process.

Echoes (or delays) and modulation effects, such as chorus, flanging effects, and pitch-changing algorithms, all come under the heading of time processors. These processors modify the time envelope of the signal by expanding its original signal and putting it into a new space, either natural or contrived, and of our choosing.

The last category of sound processors is *noise processors*, an area of signal processing and research that continues to expand and grow far beyond what was imaginable in pre-digital times. With this technology, unwanted noise can be attenuated nearly transparently, often unobtrusively, and with far greater success than has ever been possible. Background noise that might previously have ruined a good take, such as heating, ventilation, and air conditioning (HVAC) systems, wind, rain, and other constant sounds occupying many frequencies, are all easily managed, dealt with, and attenuated. Advanced noise processors can repair distorted audio to an extent and remove offending noises, such as squeaks, rattles, and even sirens and cell phones. Even unwanted reverberation can be reduced, and more advanced functions are being developed as AI becomes integrated into the technology.

Of course, there are limits to this technology, and it should never be an excuse not to strive for the best quality throughout the production. It is paramount that we take all the steps necessary to get good sound in the first place, and just because this newfound ability to make things better in post-production exists it doesn't mean we can (or should) settle for inferior audio from the field. Noise processing technology may improve bad audio, but it will always make good field audio sound even better.

1.4. DISTRIBUTION SYSTEMS AND MEDIA

There are many ways to experience media, and it's only becoming more diverse as the requirements of the varied media formats continue to evolve and grow. There are legacy media, such as theatrical film – which may or may not be on film these days – and terrestrial or broadcast television, which are being supplanted by streaming media. Other than sporting events, these may be the waning days of terrestrial "event television," where viewers plan to be in front of their television at a set time and on a set day to see a favorite show. Although linear broadcast television is almost certainly here to stay, it's no longer viewers' first choice for consuming their favorite programming. Viewers want their programming on their terms and at their convenience, and available to them on a very wide variety of devices and platforms. Our forebears in the field would have seen all this as a logistical and technological nightmare.

It is a lot to keep track of because all the potential media outlets and playback device combinations make it almost impossible to use any one approach that would make them all sound good. A program will simply never sound as good over a smartphone's speakers as it can over a good home entertainment system, and neither system will sound as good as a theatrical sound system. This is the new reality, with no straightforward path to reconciling all these formats and mixing them in the same way. What remains is to make all programming sound as good as possible on a baseline level and hope that that quality will also translate to all the other media and device combinations.

That's not to say a theatrical mix would be heard on a streaming service like Amazon Prime or Netflix. That film will have to be reworked and "downmixed" to work on the smaller format of the home television screen and sound system, where mixers adjust the mix of the film so it works in the context of the typical home viewing setup. The manufacturers of media consumption devices are also striving to make them sound as good as possible while working from the same baseline of what constitutes a good-quality sound mix. A few years ago, it would have been hard to believe that a smartphone could have decent-sounding stereo speakers, but technology advances, and now many of them do. On the higher end, soundbars and other home systems have made giant leaps forward, sounding better with every iteration while doing it more cost-effectively at the same time.

The trick comes in accommodating all potential media consumption channels and knowing the differences between them all. It comes down to mix parameters, channel distribution (perhaps stereo or some form of surround), general mix processing (compression, EQ), and then knowing what will work best in the different realms. Broadcast, or anything with a distribution life span in either streaming broadcast (Netflix, Amazon Prime, Hulu, etc.) or terrestrial broadcast on the major networks, must meet specific, stringent level specifications. Other streaming – such as YouTube or individual websites that might have their content hosted on YouTube – will typically run significantly louder than most broadcasters can handle. Theatrical – and some streamers, and certainly anything headed for DVD release – will follow yet another set of cinematic deliverable requirements, adding more sound channels to the mix, such as 5.1, 7.1, and ATMOS. Then there's crossover programming, such as releases that play in both the cinematic and the streamer realm at the same time. The sound team needs to accommodate the requirements of the theatrical space but also create a home- and streamer-friendly version of the mix because not everyone is set up to hear immersive theatrical sound formats in their living room.

It can be a lot to consider.

In the end, like so much of the work in sound, it comes down to planning and ensuring our work translates across the different combinations for playback. That program must make the same impression, regardless of format. All mixed versions of it must sound at least similar so that a viewer who sees a film in the theater can have close to the same experience in the comfort of their own home and in whatever manner they choose to enjoy it, whether it be on earbuds, headphones, TV speakers, or a nice surround or soundbar system. To that end, there is a path to repurposing the surround mix – downmixing – so that the mix will translate and read the same across different versions.

There are many ways to enjoy a soundtrack. Here is a simple table to help keep track of all the permutations.

TABLE 1.2 Soundtrack channel permutations

Mono (short for monaural)	One channel of audio program

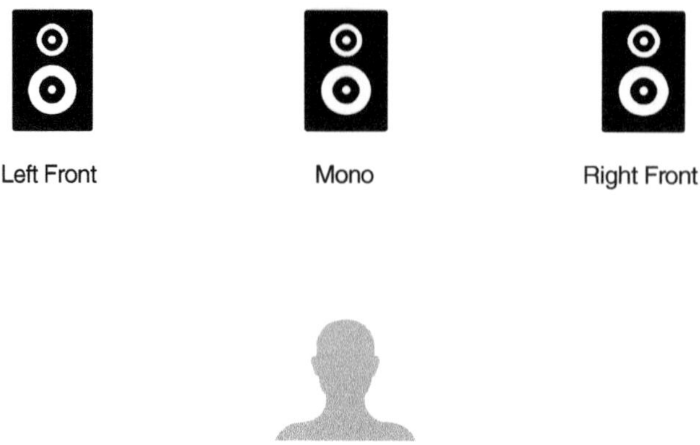

FIGURE 1.35 Stereo and mono speaker channels

Stereo (short for stereophonic) or Surround – there are many!	A stricter definition of stereo would mean two or more channels of program audio, but many times, it's assumed to be just two, in the form of left and right channels (see Figure 1.35).
LCRS, also Dolby Stereo	A total of four channels, usually encoded into a stereo pair, consisting of Left Front (L), Center Front (C), Right Front (R), and Surround (S). An encoded stereo version of an LCRS mix will be noted by the channels labeled Lt/Rt or Left Total/Right Total. The mixes are downward compatible, meaning they can be listened to in standard stereo or decoded surround modes, making them versatile and compatible with broadcast and other limited-track output situations.

Table 1.2 (Continued)

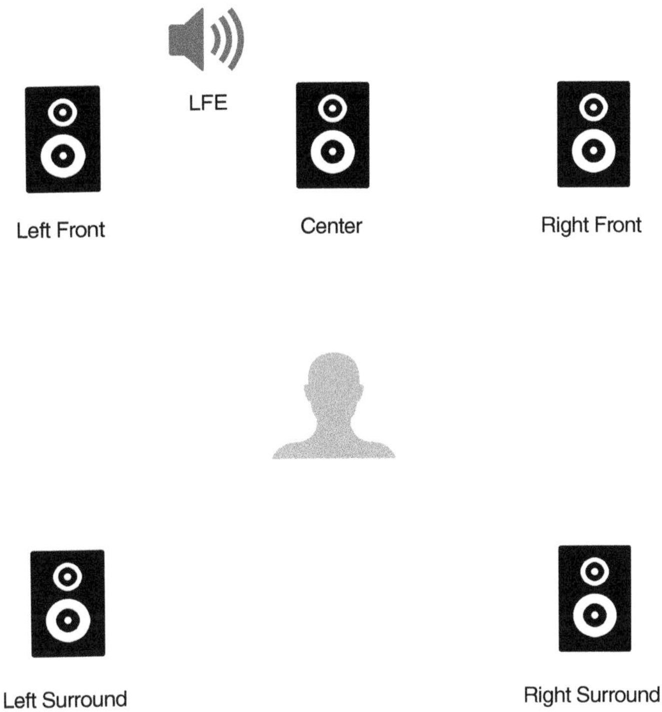

FIGURE 1.36 6-channel speaker placement for a 5.1 listening experience. Note the LFE can be located nearly anywhere, as lower bass frequencies are omnidirectional

5.1	6 total channels, consisting of Left Front (LF), Center (C), Right Front (RF), Left Surround (LS), Right Surround (RS), and LFE - also known as the subwoofer (see Figure 1.36). This is the ".1" of the 5.1, in that it is only a tenth of the bandwidth of the other channels; it's only the extremely low frequencies that reside here.
7.1	Same as 5.1, but with the addition of middle surround positions (see Figure 1.37).

(Continued)

Table 1.2 (Continued)

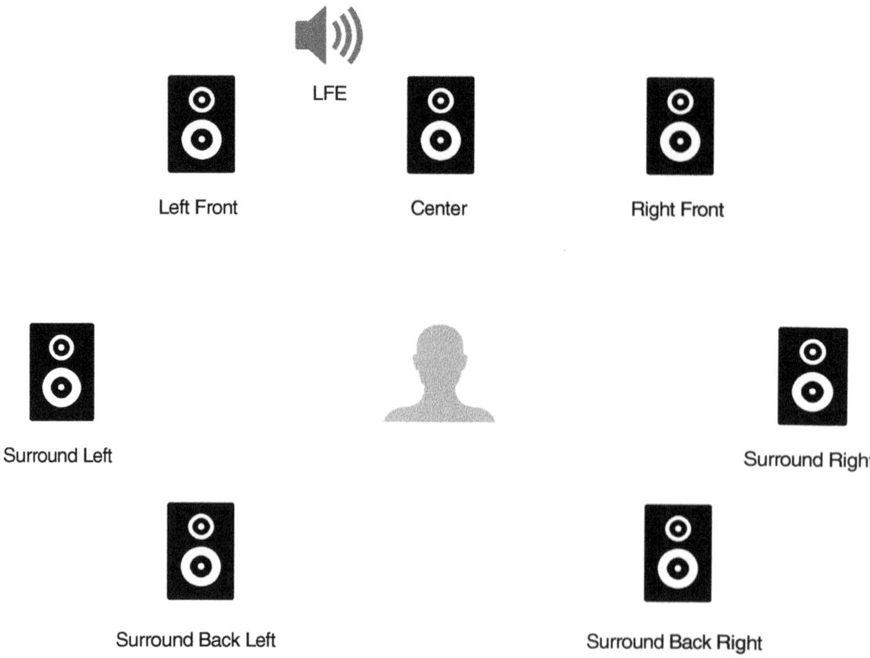

FIGURE 1.37 8-channel speaker placement for a 7.1 listening experience

ATMOS	The adaptation of ATMOS (see Figure 1.38) to home environments is different than what has become the new normal in theatrical projection. At its core, the system is designed to provide a more immersive environment than previous single-plane surround systems, like 5.1 and 7.1, by adding the vertical dimension – hence, we see new configurations like 5.1.4 (where the ".4" represents the overhead speakers) and theater formats can vary from there, from 7.2.4 on up to 11.2.6 – and more. The system's flexibility is the key, and the main point is to create the vertical axis in the sound field.

Table 1.2 (Continued)

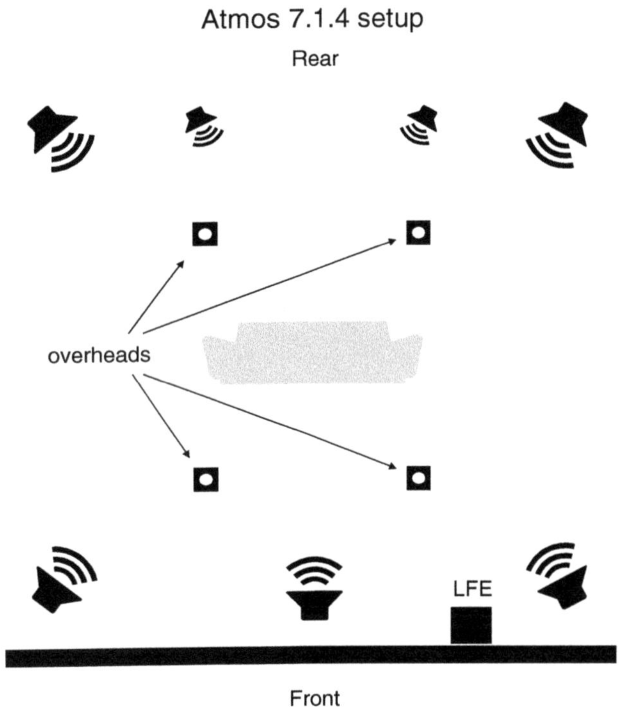

FIGURE 1.38 Typical Dolby ATMOS speaker setup – 7.1.4 array

CHAPTER TWO

People and Processes

2.1. PRODUCTION AUDIO – PERSONNEL AND HOW THE GOOD WORK GETS DONE

When looking at a typical call sheet for a day on set, the number of personnel on any given day can range from less than a dozen to literally hundreds, once all the positions and above-the-line people are accounted for. In production vernacular, *above the line* includes producers, the director, the assistant directors, and the main cast; *below the line* includes the extras, but is mainly key crew-members from all the departments and support services that may be needed on any given shoot day.

These people are all there to do the vital work of getting the day's scenes shot and in the can.

The simple truth that the sound crew must contend with is one of lopsided numbers. In the end, very few people there are specifically focused on or concerned about the quality of the sound, even on the largest productions. More than one mixer has lamented that it sometimes seems like the powers in charge would be perfectly happy to go back to the days of silent movies rather than deal with the needs of an annoying Sound Department.

It can be discouraging at times and leads to a sort of bunker mentality in the production Sound Department. There's often a feeling of "us vs. them" as the sound crew works in pursuit of clean and usable production audio. At times, it can feel like all the other crafts are even working against the best interests of the Sound Department, with the Lighting, Set Decoration, and Props Departments seemingly designing the set in such a way that it excludes any hope of gathering decent sound. Their setups and designs leave little to no room to bring in a microphone; perhaps the set is lit in such a way that it virtually guarantees a

 DOI: 10.4324/9781003373094-3

shadow from the boom pole; and it can seem like the cinematographer stages shots so that there's no room for the boom operator to be anywhere near the actors. All while, inexplicably, the Electric Department decided to park their diesel-powered generator right next to the set.

These are just some of the daily struggles the Sound Department might encounter on a film set. Sometimes, the challenges are logistical; sometimes, they're interpersonal and become political. To do its job effectively, the Sound Department needs to overcome all those challenges somehow, while at the same time remaining singular in its purpose … but in the most diplomatic way, without seeming obsessive or acting in complete disregard to the needs of the other crafts. To do their job effectively, the Sound Department will need some help, and they'll need to foster partnerships and gain some cooperation from other departments on that set to be successful. This is where a little outreach, ambassadorship, and diplomacy can go a long way.

Looking at a typical feature film call sheet, the picture personnel-to-sound personnel ratio can be anywhere from 10 to 1 to even as high as 100 to 1 (or more) when comparing the number of people who are there for something other than sound versus those who care most deeply about sound quality. Most of the crew are on set to optimize the visual elements, such as lights, camera, set dec, props, and wardrobe – the list goes on.

The Sound Department is the only group on a set whose sole responsibility is to ensure it all sounds good. Ultimately, the Sound Department's interactions with the other departments will be the key to their success – or failure. When interdepartmental cooperation doesn't happen, it can be challenging for the Sound Department to be very effective. This isn't to lay blame on those other departments; getting the shot is their singular purpose and priority on set at any given moment, just as getting clean sound is ours. However, it's clear to see how this compartmentalized view can make it very easy for the sound crew to fall into an underdog mode of thinking. All those other departments will work to ensure their own interests and mission are covered on set, just like the Sound Department is there solely to gather clean and clear audio. The trick is to find some common ground and work with those other departments while staying out of the way.

At least three people can be counted on to care about the sound quality on a set – the mixer, the boom operator, and the third or utility position. This doesn't seem like enough, given that sound conveys at least half the story – without dialogue, establishing a narrative can be difficult, to say the very least. The hope is that others might also care deeply about it, but this is not usually the case. This calculus leaves out the most important position, however, and in addition to the Sound Department, the director should be the biggest champion for achieving great location sound.

Dialogue is a key component in narrative and documentary filmmaking, and the viewer must be able to hear the spoken word to understand what's going on. While the visual element is essential, it's important to remember that it isn't the whole story. Sound is fundamental to good storytelling; the viewer simply isn't experiencing the full story if they're only seeing the picture, any more than if they're hearing the dialogue track by itself without the picture. If, sidelined by the pursuit of stunning visuals, the dialogue track isn't up to the task of carrying its part of the narrative, then the sound team has failed in its mission. If the set culture is such that the dialogue has gone from being a story-defining element to a practical annoyance on set or even a distraction, then all that good storytelling work is undone before it can even begin.

It becomes easy to see how a set can become political, even tribal, at times. The highest-functioning sound departments, therefore, need to operate on a more holistic, cooperative, and interactive level with those other departments, and it all begins with simply being nice. The Sound Department will likely need help from several different departments at some point, so they must get on the same page as everyone else and stay ahead of the situation as much as possible.

As the head of the Sound Department, the sound mixer usually has their hands full, so on-set outreach and diplomacy will often fall to the boom op and utility positions. These two people are often in the middle of the action on set and dealing directly with the actors and the director, so as the outward-facing members of the sound crew, they're also in the best position to find allies in other departments. They're effectively the ambassadors of the Sound Department, helping to head off issues with some proactive diplomacy.

Sound will always need help from the Electric Department, certainly to get power to the sound cart, but also to convince them that they should consider relocating that noisy generator farther from the set. Sound may also need to ask the Lighting Department to redirect a light or at least put up a flag to keep the boom pole from casting shadows in the shot. The boom op should try to be on or near the set during preliminary lighting, making friends, and, quite frankly, defending what sound turf they can. Sound wants to stay in the good graces of the Set Dec and Props Departments, as their help might be required to quiet or remove noisy props. Sound might need a quick consult with Locations for help with shutting down noisy systems or even finding a place close to the set for the sound cart.

Props, Set Dec, and Wardrobe should all be in alignment with the Sound Department's mission, and endless scenarios help make that point. Sound might ask Set Dec to replace or remove plates and cutlery when they're not being seen on camera because those items will make noise that will compete with dialogue. Props might be able to help by replacing those noisy ceramic plates with paper once they're out of the shot. The leather dress shoes an actor is wearing might need to be dressed or treated to be less noisy.

The bottom line is that Sound will always need some help, perhaps even a lot of help, to be effective at their work. These transactions tend to be very one-sided, as there's very little the Sound Department has to offer other departments.

To help initiate these transactions, some sound departments keep a spare set of headphones (or two or three) at the cart position for anyone hanging around to listen in on the dialogue as takes are being recorded. This is part of the outreach Sound can do, and it's often an eye-opening (and ear-opening!) experience for crew members from those other crafts who might not otherwise have an opportunity to hear the production track. Despite being close to the set, many crew members will never hear the dialogue until the project sees release. Even if they're close to the action, they're not hearing it in its best, most pristine state; they are completely removed from that aspect of the shoot. Offering a set of headphones is an easy bit of outreach that's simple to do and is a small dose of goodwill that can sometimes go a long way toward getting people to understand and get on board with the sound mission at hand. Some mixers are not above leaving a spare phone charger or two plugged in and available for general use, or maybe even putting out a bowl of treats for other crew passing by.

Ultimately, film production is very much a team sport; it needs to be completely collaborative, or it will fall apart. There's no harm (and a world of good) that will come from little reminders to the rest of the crew about the sound team's contribution to the finished whole. Providing access while avoiding needless "turf wars" will go a long way toward generating a collaborative, productive, and creative set culture.

2.2. PERSONNEL

Starting at the top of the Sound Department, there's the position of *production mixer.* This is also the department head, directly answering to the directorial staff while ensuring that all sound gets recorded properly and as close to perfect as possible. As one might guess, the job requires much more knowledge and expertise than simply pressing the *record* button; the mixer is working to ensure that all sound produced during the shoot is viable, clean, and entirely usable.

"Our primary goal is to capture the dialogue as it is written on the script," says Tom Williams, CAS, a feature film mixer whose credits include *Black Mass, Don't Look Up,* and *Free Guy,* among many others.

It would seem to be a simple enough task, but many things can interfere with the process – some of them expected, but so many more that are hard (even impossible) to foresee or overcome. Nevertheless, the mixer must try to predict these challenges ahead of time, and it all begins when they first lay eyes on the script, which happens weeks or sometimes even months ahead of shooting. The script will likely evolve, occasionally even during production, so there is a constant need for assessment. This process continues with the arrival of each day's sides, which are the individual pages of script that will be covered in a given shoot day.

"It's all about making the set as quiet as possible," Williams says. "If it's cars running or compressors running or generators running or whatever. It's just mitigating and knocking down noises so we can capture that dialogue clean. That's pretty much our goal."

The production sound mixer has their own blocking process, where they figure out the needs of the production, including what microphones will be needed in certain places and situations, while also anticipating any noise issues that might conceivably pop up and the possible mitigations for those issues when – not if – they do.

"There's homework involved ahead of time for sure," says Tom Williams. "I read the script as a story at least once, if not twice, then I'll start reading it again specifically to tear it apart for specific sound issues. What you want to do is make your notes before you talk to the director so you can offer up as many options as is necessary."

In addition to ensuring that all takes get recorded cleanly and at a good level, each shoot day's takes need to be laid out logically, labeled, and presented in a well-organized form so they can then be shipped or copied off correctly for the post-production team later. It can be a lot to manage, and all of it is the responsibility of the mixer.

The *boom operator*, also known as the boom op, is the other end of the sound-gathering equation; they are the tip of the Sound Department sword, in a way, and will always be on set and in the middle of the action, interacting closely with both the talent and the director (and assistant director, camera, and lights personnel) as well. The boom op is there to ensure the mixer has good sound to record, but they're also there to stick up for the Sound Department's turf. When takes aren't rolling, they will work with the rest of the crew to ensure they still have room to operate on set and won't get lit out of a scene because shadows can quickly reveal a mic boom's location. Visually reflective surfaces, like television and computer screens, windows, and, of course, mirrors, can make it more difficult for the boom operator to do their job effectively because they eliminate a lot of places where the boom operator can hide on or near the set. The last thing anyone wants to see is a full-on reflection of the boom op in the shot. Having the boom op around the set during pre-light is a big help if the timing allows for it, and their presence further helps enrich the collaborative nature of the interaction of the Sound Department with the rest of the production.

The *third* – sometimes called the cable person or the sound utility position – is too often an overlooked or unbudgeted position in the Sound Department. A good third is essential to the smooth operation of the Sound Department, however, and fills many roles that make the department more efficient. They have a thorough knowledge of the sound cart and all its contents, and, even more importantly, the application and implementation of all those systems. They're responsible for making sure that all wireless microphone and communication systems are transmitting and receiving properly; all batteries are fresh, charged, and ready to go; any cables that need to be run are neatly dressed off; and that all the sound gear is neatly stowed, readily accessible, and fully functional.

They can also be available to fill in as a second boom operator if needed, and they will know the mixer's sound cart well enough to help with any tasks required from that point as well.

In earlier times, having a third in the Sound Department was a luxury and might only have happened on major films with bigger budgets. In those days, the position was more often referred to as the cable person, with the primary task being to ensure all the microphones were connected properly and that the boom operator and other set personnel didn't trip over the Sound Department's cable runs. But the role and scope of the sound crew have evolved and grown over the years, with one of the most significant changes being that most of our microphones are wireless now, so it has become less about running cables and more about the many other technical tasks necessary to keep the Sound Department functional.

In addition to all their technical expertise, the third is often tasked with the administrative duties the department requires on set: keeping track of and submitting timesheets, disseminating information and safety information from the production, keeping track of the day's sides, and more. As such, a qualified third has become essential to any production for the support they provide.

For the Sound Department to function as a unit, they need to be in constant communication with each other throughout. "We have to mine our own information wherever we can find it because it's not always coming from above," says Tom Williams. "You get into bigger productions, where you have 200 people working, and communication isn't always perfect." The boom operator and the third can help with that communication, being in the positions they're in.

"When I run the sound department, we have an open communication," Tom Williams says. "There's no hierarchy. I am the titular boss, but it comes down to all those information streams coming from everybody. My boom op certainly has the stream coming from the set where they could tell me something I didn't hear, like, 'We're going to dump this scene and jump to something else.'" This is all actionable information that might not have gotten to the mixer as quickly otherwise. Williams adds, "My third, who is out there wiring people with mics, is also going to be dealing with wardrobe. So that's a whole different stream – wardrobe knows things that we don't; sometimes, they may get a heads-up days in advance. It's using every stream possible for any information that will help us be more efficient."

The size of the Sound Department will be dictated by the needs of the production, for the most part, but the demands of the budget also have some say as to the number of personnel that can be on set. Larger productions might require bigger sound departments, with a second or even third boom being needed in some cases, in addition to the sound utility position. Those bigger productions might also require a second unit. In those cases, an additional crew will be added to the rolls, duplicating the tasks in the first unit with the same positions at mixer, boom op, and utility. This second unit will work concurrently with the primary production but at different locations.

Some productions occasionally need to bring on a fourth position when *audio playback* is required. Playback capability is necessary when the actors on set need to act or interact with a music track or another sound element. For more straightforward playback situations – for example, just playing a song or a part of a song to motivate on-camera action – the mixer can usually play a track from a tablet or laptop computer, sending it out to the actors on set over speakers, or sometimes via in-ear monitors. When the situation gets more complex, such as when actors or extras are singing and dancing, or if multiple music cues or different cueing points are needed, the mixer will call for a playback person to be brought in.

Having playback complicates the process, at times, since the music track needs to be heard by the actors, but at the same time, kept out of the environment and, most critically, out of the actors' mics. There are several ways to accomplish this, starting from the low-tech "play-and-duck" method, where a track is played over a PA system and then turned down just before the dialogue begins. There are also higher-tech methods, such as in-ear wireless monitoring systems, known colloquially as *earwigs*. Playback specialists may also utilize a subwoofer – also known as a *thumper* on set – to get the beat and tempo out to the actors. A thumper lets everyone hear and feel the tempo of the music to help with the timing of any dialogue or singing, but it's at a low enough frequency that it can be filtered out at the recording, so it isn't printed with the production audio.

2.3. EQUIPMENT

The more complicated sound needs become, the more the sound equipment needs to rise to the occasion. Whereas earlier productions utilized reel-to-reel decks that recorded only one channel, or track, of audio, multitrack field recorders now provide the means to record several microphones separately while simultaneously recording a mixdown of those same microphones. Consequently, the mixing equipment that is a part of these recording rigs has become quite a bit more complicated over the years, with all the related connectivity solutions rising to the challenge. Wireless and Ethernet interfaces are now being utilized in some location systems, as well as some microphone systems. At the same time, increasingly constrained bandwidth is hitting the industry hard and at a time when it's needed most. Telecoms, consumer electronics, and emergency-band communications systems have all taken a substantial bite out of domestic wireless capacity in recent years, and its impact is felt daily on set as sound crews spend too much of their day looking for a clean part of the radio spectrum. The issue is only getting worse as the availability of usable spectrum only continues to get tighter.

In addition to the Sound Department's needs, many other systems on set are wireless, starting with the communications headsets the director and others need to hear the takes as they're being recorded. All the communications going on between all departments (typically radio communications, but also data) are wireless, but there's also a need for Wi-Fi and a viable connection to the internet – and the biggest consumer of bandwidth is picture. All these demands further restrict the range of usable frequencies available on any given set. If that set is in an urban area, those constraints become even more of an issue as the competition from the environment steps up.

The rest of the gear that a mixer and boom operator will need on set is presented here in somewhat general terms, with little in terms of brand specificity. It is all fairly scalable, depending on a production's needs, and can get as complicated (and expensive) as circumstances and production budgets allow.

First, the multitrack recorder is the centerpiece of any field recording system (see Figures 2.1–2.3). Field recorders come in a wide range of price points and functionality, making it possible to outfit a sound cart with a solution nearly tailor-made for almost any situation. Several handheld recorders have some

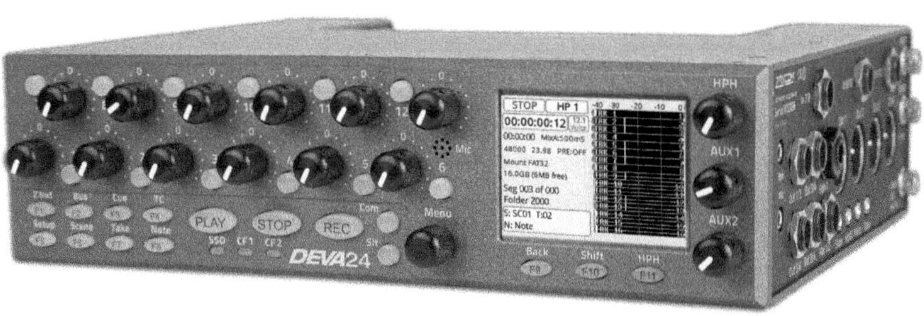

FIGURE 2.1 Zaxcom DEVA 24 field recorder

Source: Photo courtesy Zaxcom.

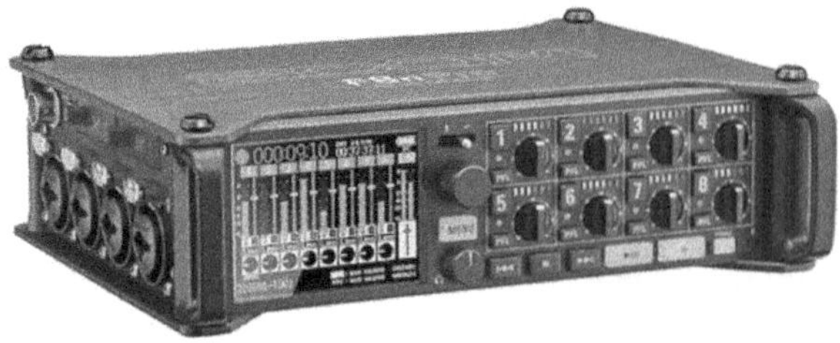

FIGURE 2.2 Zoom F8n Pro field recorder

Source: Permission Zoom, Inc.

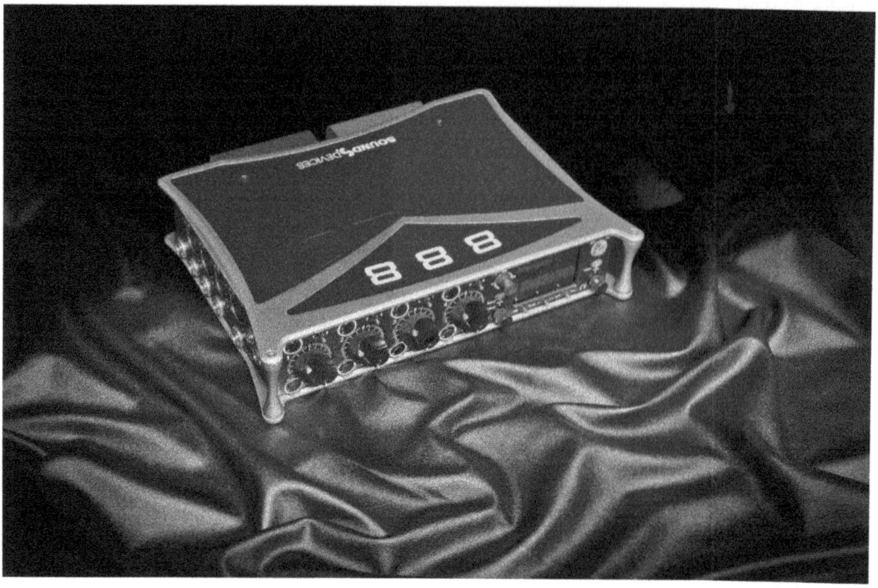

FIGURE 2.3 Sound Devices 888 field recorder

Source: Permission Sound Devices, LLC.

capacity for more professional microphone inputs, but there may be a sacrifice in microphone input quality in such a small unit. Generally, field audio decks start at 2 or 4 tracks and can go up to 24. Sometimes, multiple decks can be chained together when even more capacity and flexibility are needed. As the central element of any sound cart, it is also one of the cart's most expensive components.

In some run-and-gun situations, such as documentary, electronic news gathering (or ENG), and lighter film applications that don't have the benefit of a feature film crew, one might hear that the mixer is "going with a bag rig." This only means that it won't be practical to have a full sound cart on set and that they will work with a smaller, lighter, and more portable version of their sound

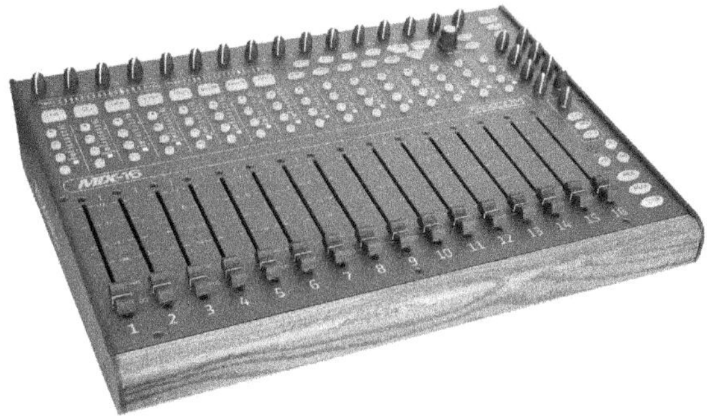

FIGURE 2.4 Zaxcom Mix-16 control surface

Source: Photo courtesy Zaxcom.

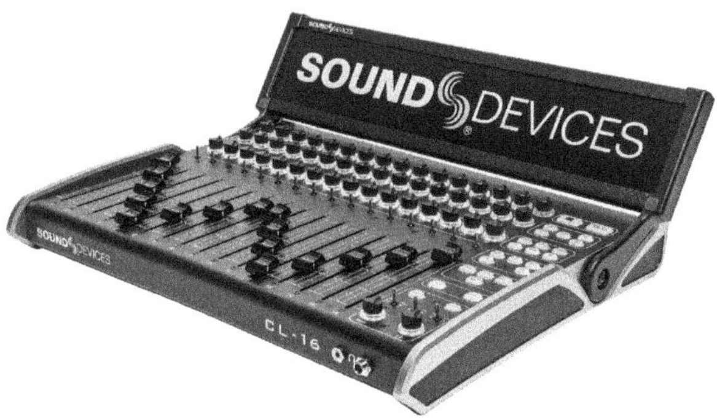

FIGURE 2.5 Sound Devices CL-16 control surface

Source: Permission Sound Devices, LLC.

rig. It usually means a smaller track count version of a multitrack recorder, carried in a portable bag, along with a somewhat stripped-down package of microphones, cables, and other essentials they may need to get through a particular shoot. A typical bag rig can be worn over the shoulder, with many purpose-built bags having built-in access points to the inputs and outputs of the recorder, along with easy access to the control panel.

Next up would be the mixer interface, or mixing control surface (see Figures 2.4 and 2.5), so-called because current-generation hardware audio mixers likely do not have an actual audio signal running through them. A modern digital mixing surface is essentially a data controller, providing the sound mixer (the person) with a user interface to more immediately access less-accessible features and controls within a sound recorder. As mentioned, it helps to think of the mixer (the device) as an extremely high-functioning controller, much like a complicated, multifeatured mouse. These are streamlined versions of the mixing surface that one might find in a recording studio but which have been optimized to fit into a more compact package for field recording, with fader controls and other mission-critical features ergonomically placed for ease of use. Controls that aren't needed as often can be accessed via the multitrack deck interface menus. These can get quite deep, as one might imagine, so the operator must be familiar with the many functions of the controller before getting into an actual working situation on set. Deeper functions require a greater depth of knowledge, and it can take some time before the necessary muscle memory and intuition develop to reach a comfort level with the equipment.

The next set of mission-critical gear is the microphones themselves, and delivery systems are needed to get them into the mixer and recorder at the sound cart. Typically, this work is done wirelessly, but any Sound Department should also have an ample supply of microphone cables of all lengths in case bandwidth and radio interference (technically Radio Frequency Interference, or *RFI*) becomes an issue on set. While it may be true that the most secure and surest means of delivery will always be over a length of high-quality cable, the needs of utility, convenience, and portability dictate that most microphones used on a set do need to work wirelessly.

The three components required to run a wireless microphone setup include the microphone itself; a high-quality transmitter in a small, battery-operated pack that the microphone plugs directly into; and the receiver, which connects directly to the recorder input. A preamplifier in the transmitter is designed to take the very low output level of the microphone and amplify it to line level, setting up a better and cleaner transmission for ease of recording and any manipulation of the signal that needs to happen along the way.

The microphones can be very task-specific, each designed for situation-specific needs and applications. Just as not every tool in a toolbox is a hammer, there are many times when just one type of microphone won't work for every situation. On set, our primary microphones fall into one of two main categories: either mounted at the end of a boom pole (referred to collectively as "boom mics"), or lavalier-type mics meant to be physically placed on the person being recorded.

The preferred means of sound gathering will nearly always be a boom microphone (see Figure 2.6), as it has the best and most realistic sound qualities. These are often described as having an open feel that provides perspective. The term is generic, as many different microphones fit this description and application. Another commonly used term is "shotgun," a generic term for any directional microphone.

The other type of microphone most used on set is the lavaliere microphone (see Figure 2.7), an all-encompassing term that applies to a whole class of miniature microphones. Their main strength is their minimal size, making it easy to hide a lav on a person's body or tucked into clothing. These microphones work

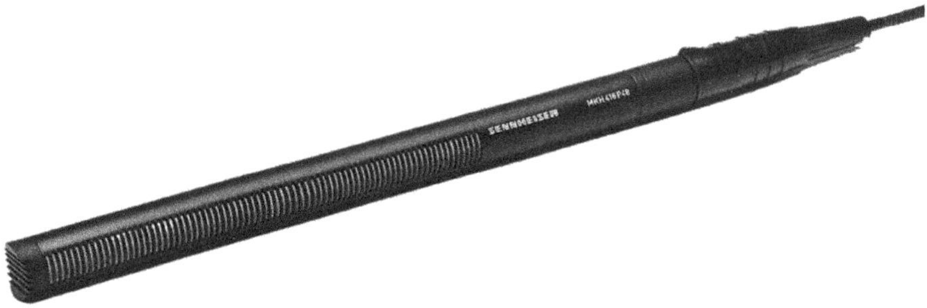

FIGURE 2.6 Sennheiser MKH 416 boom microphone

Source: Photo courtesy Sennheiser.

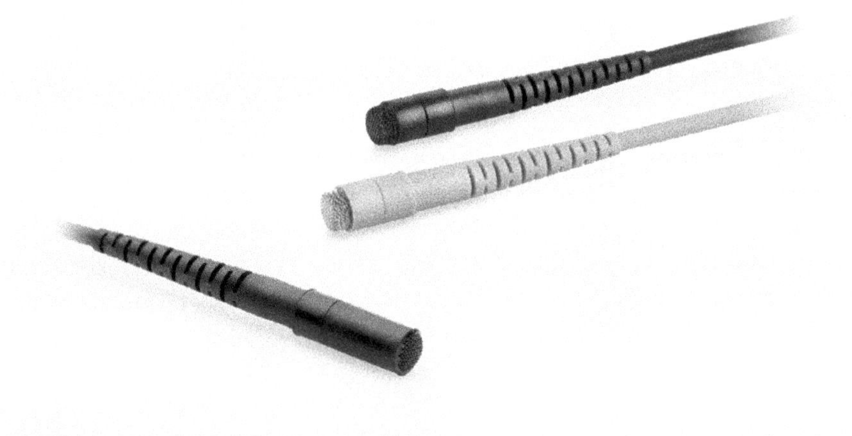

FIGURE 2.7 DPA 4660 lavaliere microphone

Source: Permission DPA Microphones, Inc.

well and are small enough to be nearly invisible on camera, making them the ideal solution for getting clean, accurate sound.

Despite the technological leaps, the preferred method among sound professionals for gathering sound on set, particularly dialogue, is the boom-mounted microphone. By necessity, the boom mic will always be at some distance from its subject; it can never be closer than the camera frame's edge, but this also helps it provide a more realistic perspective. There will always be some distance between the microphone and the subject, so it also captures a broader snapshot of the source of that voice, resulting in a much more realistic representation of the human voice in the end.

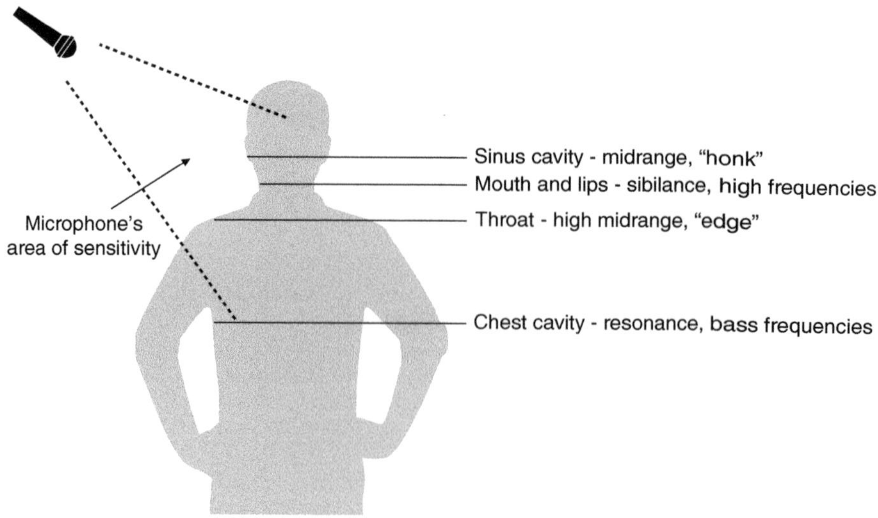

FIGURE 2.8 The human body's frequency ranges by region

When recording the human voice – or anything, for that matter – it's essential to consider where the sound is coming from. In the case of humans, it emanates from several places in the body, not just the mouth (see Figure 2.8). The tongue and lips help form and define the sounds we know as language, so it may seem most obvious to just point a microphone at them, but humans don't listen to each other with their ears positioned directly in front of the mouth, making this approach somewhat unrealistic as a method for recording voice as well.

Consider all the places in the body that are involved when speaking. The sinus cavity contributes some midrange frequency, sometimes manifesting as the honking quality in the voice when a person has a cold. The throat can give some rasp or edge to the vocal quality, and the chest resonates with the lower bass frequencies. Capturing this whole picture, instead of just the area around the mouth, will provide a more holistic and realistic representation of what that person sounds like, in part because it emulates how that voice would be perceived in the real world.

This is why the boom mic, with its mandatory distance, sounds better and more realistic.

While convenient and able to be located right at the sound source, the lavaliere microphone lacks all that perspective. It is also susceptible to interference from wardrobe issues and the physical actions of the actor wearing it, as it needs to be stashed out of sight of the camera while at the same time being in a place that will pick up the voice most effectively. There are formidable factors when gathering clean dialogue, with the expense being another factor. The best gear is also usually expensive, but the good news is that high-quality microphones are more robust and somewhat futureproof so that they can provide many years of service with proper care.

With different transmitter/receiver combinations, there are also issues with whitespace restrictions or the lack of bandwidth available. Different regions and countries have different frequency bands available for public use, so be sure that the gear being brought to set is even allowed in the area where it will be used. In many countries, a level of federal oversight comes into play, and those authorities will impose fines if they detect equipment being used outside of the publicly accessible frequency band ranges.

There are accessories and other pieces of equipment that can help make the sound mixer's job easier, but before any sound gear is installed, there is the power management issue. For any mixer, but especially one who specializes in working on bigger and busier sets, there will need to be a battery backup system on their sound cart. These can be heavy and expensive, but they're also virtually a requirement on a busy set, particularly if that set is in remote locations.

Time is money on a film set; sometimes, there just isn't time for the Electric Department to get power drops out to all the departments before the crew is expected to be up and running and ready to roll. Consequently, onboard power systems must be robust enough to run the entire sound cart and all its essential peripheral gear for at least a few hours. Fortunately, there has been some standardization over the years, and most of the gear that one would put into a location sound cart operates at a standard of 12 volts DC.

Aside from the mixing interface, the recorder, and the wireless receivers, the sound mixer on a larger shoot will also require video monitoring capability so they see both the framing of the shot and the action on set. A movie set is a busy, crowded place, so the mixer and their sound cart often end up situated far enough away that they won't be able to see the action directly. The mixer typically communicates with the boom operator directly to let them know when the mic is in danger of dipping into the frame, but most importantly, they need to see the action on set to help them anticipate the line delivery from the actors so they can have the right microphones open at the right time.

Sound departments use a closed communications system, commonly known as an Interruptible Foldback system, or *IFB*, to communicate with their team directly and across those distances. For instance, it can be helpful to warn the boom op when the camera operator tilts up or goes wider than expected, and so there's now a greater risk of seeing the boom mic in the frame or its shadow. It can be helpful for the boom op to communicate any last-minute blocking or line changes happening in the upcoming take. The boom operator and mixer can also better coordinate their moves with each other logistically; for example, the boom operator might not be able to get a microphone over in time to "reach" an actor's entrance without either entering the shot or casting shadows, so the mixer must know about that ahead of time. The mixer can then know they will have to have that actor's lavaliere mic on and ready for a particular line before they "toss" back to the boom operator to cover the rest of the scene. When combined with the video feed, this back-and-forth communication helps the boom op serve as the mixer's eyes and ears on set, which, in turn, will help them to mix the scene better because they can better anticipate the action as it's happening.

The modern sound cart is typically kitted out with a large variety of equipment and solutions – they've got everything but the proverbial kitchen sink (see Figure 2.9). A well-designed cart's strengths quickly become apparent, particularly when working on bigger productions, especially narrative features. It can be a lot to haul around, but having a loaded sound cart helps the sound team deal with less-than-optimal conditions and gives them the flexibility to handle any number of conditions and circumstances effectively.

Of course, there is a limit to how much equipment can be wedged into the confines of the sound cart platform, so the sound team will also employ a second or follow cart with any extra gear they might need (Figure 2.10). Often built on a standard, heavy-duty AV-style utility cart, it must be big enough to hold a

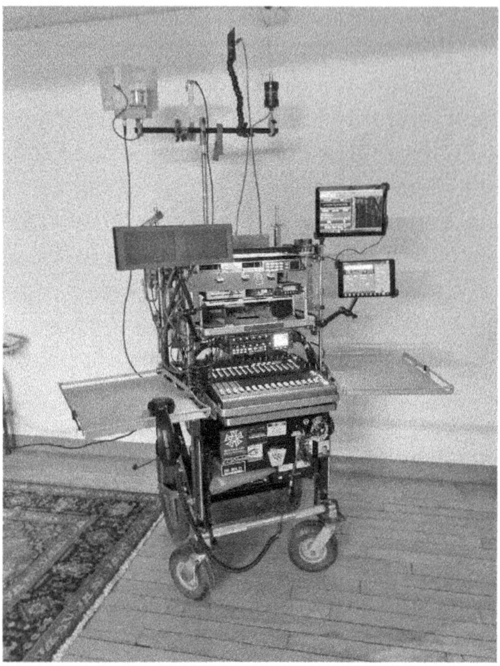

FIGURE 2.9 Tom Williams soundcart

Source: Courtesy Tom Williams, CAS.

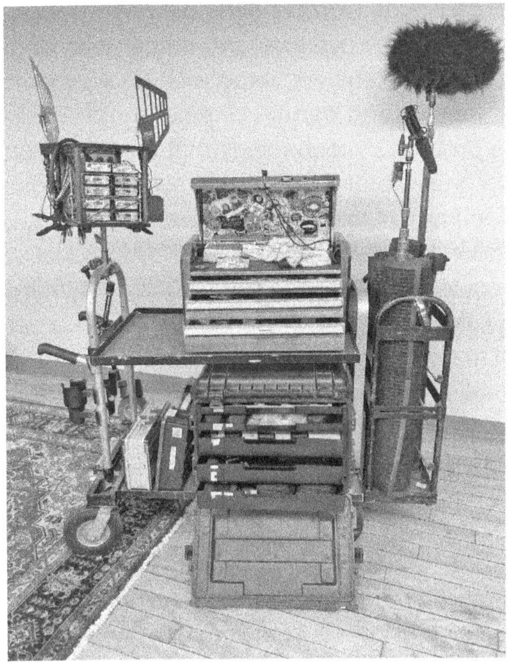

FIGURE 2.10 Tom Williams follow cart

Source: Permission Tom Williams, CAS.

wide range of equipment and the supplies required for the production at hand. This will include organizing drawers and bins filled with all manner of batteries and their chargers, plus an array of microphones, comsets, connectors, cables, and adapters, along with a long list of expendables and other accessories – sometimes referred to as gags – that are part of the more personalized part of a mixer's kit. One will often find an odd collection of items appropriated from a variety of sources. Everything from cable ties to toupee tape, surgical tape, various types of foam, cable ties, and sticky putty, this list can go on and will always vary and be highly personalized to the mixer putting it together. The range and type of items can be very proprietary and individualized, but all of it is there to help with the work of attaching and stashing microphones, either on set pieces or on the actors themselves. There will also be a supply of microphone accessories, including holders, windscreens, extra boom poles, various shock mounts, and lavaliere clips and mounts for various applications. Everything on the follow cart is there to get those microphones optimally placed, secured, and as close to the action as possible.

Somewhat unique to the Sound Department, and location audio in general, is that the sound mixer typically owns most of the audio gear used on set instead of just renting it. Comparisons between Sound and some other departments can in this way overlook that some gear, especially lights and camera lenses, can be prohibitively expensive for an individual to own. Further, many lights, lenses, and certain cinema-format cameras can only be leased, as they're proprietary technology that simply isn't for sale.

Sound people prefer to bring their own setups and gear for expediency and utility. The economics of the equipment are more conducive to ownership, and the ability to tailor one's workflow to expedite individual styles of working is part of a mixer's reputation as a professional. It's part of their personal and professional identity, refined and developed through hard-earned experience and over the course of many productions. As the owner of a particular kit, the mixer will then rent it to the production company for the length of the production, so it also becomes a critical part of their revenue stream. A mixer will usually set up a business identity that technically owns the gear and will then create an individualized package from the company's inventory that best suits the needs of the film at hand, invoicing the production as they go.

A feature film mixer will be prepared for anything; consequently, their equipment inventory is large enough that it often requires a truck or trailer space – that extensive collection of gear isn't always directly present on set. A feature film mixer is expected to have all that equipment available at a moment's notice, however, with the understanding that some of it might not be needed very often – if at all. In any case, the mixer still needs to be able to meet any challenge or need as it happens in real time.

The fundamentals of the kit remain the same, however, from feature to documentary to independent projects, and what any mixer needs foremost is a recorder, some microphones, some transmitters and receivers, a boom pole, some headphones for monitoring, and all the necessary cabling to make it all work together. Any additional accessories are up to the individual and the needs of the production.

The recorder is essential, and the technology is similar across the current range of decks available. The primary differences lie in the quality of the inputs, the durability, and the ergonomics of individual machines, and then it comes down to choosing the deck that will work best for a particular application, as well as the personal preferences of the mixer using it.

Tom Williams says the most important gear in any production sound setup is the microphones a mixer chooses. "Good wireless systems and good lavalieres are very important, and probably even more important than the recorder." Technology has made considerable advances in this area over the past few years, and the latest lavalieres sound much better than even ten-year-old technology.

Tom Williams got his start in documentary filmmaking, which taught him many lessons about being ready for anything. "It's all about being prepared;

that's the biggest thing. You don't win an Academy Award because you use the fanciest gear. You win it because you do the work, and the work is being prepared. All that other stuff will follow – my big thing is, technology isn't always the answer."

2.4. BEST PRACTICES IN THE FIELD

So now that we've got the gear and our marching orders, what happens next?

The sound crew's goal is to do its work as quickly and unobtrusively as possible, and that work should not call attention to itself. The last thing any sound person wants is to slow down the production's workflow; the very last thing we want to hear is the assistant director shouting, "Holding for sound!" while we try to get our collective act together. Again, the call sheet for a typical production shows dozens of people on set to make sure everything *looks* great, most especially the lights, camera, grip, electric, props, set dec, wardrobe, and makeup departments – to name some – but of those dozens of personnel, it's the sound team that ensures it sounds great as well.

Of course, experience plays a role in anticipating what setups might be necessary well before they're needed, yet at the same time, the sound crew needs to be ready for almost any eventuality. To prepare, the mixer will spend some time breaking down the script, making informed decisions and educated guesses.

As just one example, a scene might call for a wide shot, but the lighting and angles have all but eliminated the possibility of using a boom microphone. But after getting that establishing shot, the plan is to move into tighter shots, framed in such a way that we can probably get a boom mic in there after all. The mixer needs to decide if it will work to have the boom a little farther off-set for that long shot, just so they can be consistent with the sound mix quality when the shoot goes to the tighter shots, or isos. Having an open boom mic near the action might help make that sonic blend of microphones more consistent across all the angles later in post, making life easier for the dialogue editing team when their turn comes.

The mixer will also have to consider the number of principals who have lines to deliver, what props or vehicles might be used during the scene, and the general environment of the location. All of these will influence (if not directly dictate) how we handle the scene, sound-wise, while at the same time limiting many of the options we might have to get that dialogue recorded. The mixer will then make the call, selecting the equipment to have on hand for the day of the shoot while specifying some additional pieces as alternatives and backup, just in case.

Sometimes, problems arise when the Sound Department is *too* efficient. The best sound crews are adept at predicting the future and providing the gear and the setups needed on set before shooting happens. Still, even the best directors can get used to that level of preparation and preparedness and begin to take it for granted. It can then be a little bit of a shock when the unexpected happens, and a situation arises that the sound team wasn't prepared for – the one constant in the industry is its ability to surprise. Just when the Sound Department might be starting to feel comfortable or even certain that they have everything locked down and are ready for anything, the inevitable surprise pops up, and they will have to respond to that situation as quickly as possible.

Sometimes, it's beyond any individual's control, such as a sudden weather change, power issues, equipment failures, or unexpected and unwanted noise on set. The Sound Department can only respond to the extent of their abilities and control. Sometimes, it's much smaller in scale, like an actor missing their mark, a background actor being given a last-minute speaking part on the third take of a scene, a camera move not executed properly, or a lens change happening unannounced and suddenly putting the boom in the frame. These situations are even more annoying because they could have been easily managed with transparency and communication. If someone had thought to warn the sound team that, this time, that one background actor would be delivering a new line at the top of the scene, that information could have been relayed to the boom op on set so they could get the mic over to that actor in time to catch the line or, if they felt they couldn't make it, the utility person could have had that person wired in under a minute.

These communications typically fall to the assistant director, who is out there to manage the set and keep the sound crew up to speed. Sound crews also try to be extra vigilant and head these things off as they happen in real time, especially the boom operator, who is often closest to the action. The best boom operators are eavesdropping on these conversations as they happen on set. They will know what's happening before anyone else and then relay that information back to the mixer. Boom operators and mixers need to know how to read lenses to recognize a lens change as it's happening, noting whether it's a wider or narrower field of view. At this point, the boom op will check in with the camera operator while the mixer refers to their monitor to relay to the boom op where the new frame edge is in this new shot.

Ultimately, it comes down to communication and facilitating the flow of information. The better that information flows, the better the result will be. Some would say this is a cultural thing, and every set develops its own identity – a set is a living, breathing entity that develops its own culture and identity. When the Sound Department does so much of its work under the radar, being out of sight so often means out of mind. But just because a Sound Department is adept at anticipating the needs of the production, it doesn't mean that they can't be caught off guard without the right inputs and direction. These are the times when the sound team is cursed by their successes, where successfully anticipating 99% of a production's requirements sometimes means that the leaders of that set have essentially forgotten about them, leaving them out of the loop when a crucial change occurs. Being out of the loop can be detrimental to getting good sound and keeping the forward momentum going on set.

As mentioned, it can feel very "us vs. them" when these situations occur. That train of thought is only human nature, and it happens pretty easily, but it's important not to go down that path and fall into this bunker mentality. It's important to remember that working on a film is a collective exercise that works best when viewed as a team sport rather than as a solo undertaking. If any of this was easy, anyone could arm themselves with a camera, some lights, a microphone, some actors, and a location and easily create compelling – even award-winning – film stories on their own. It will always make for a better result when specialists' creativity and technical know-how are effectively harnessed to get the job well done.

The main objective of the sound professionals on set is to follow Tom Williams' advice and get clean dialogue before all else. Everything else is secondary, and so everything else can wait. Cool, bespoke sounds that are unique to the location or environment or props and vehicles being used can be recorded, of course, but it shouldn't be at the expense of the workflow or the process of the entire production, and never at the cost of getting the best-quality dialogue recording possible at that moment.

This is undoubtedly the case when working within a narrative story structure, where the storyline and production workflow are scripted and tightly controlled, at least to some extent. Documentary production is much more "of the moment," and the sound crew needs to gather all the local sound they can while there in that location. Many remote or specialized locations will be much harder to recreate sonically after the fact, and the production would be sacrificing some of its integrity to have manufactured sound backgrounds placed in the final film instead. Most importantly, interview subjects will likely not be available or even willing to come back after the fact to redo an interview, nor should they. That interview was a moment in time and place, and as a verité work, it would compromise the story's integrity to have an interviewee recreate any part of it.

On the set of a narrative film, the mixer might have some ability to acquire those more bespoke sounds in the moment and in ways that don't intrude on that edict of getting clean dialogue. While working on the set of *The Town*, Tom Williams had an opportunity to plant microphones in a parking lot full of parked cars that were about to be crushed by an out-of-control armored truck. That would be a unique sound, and even though that sound could easily be designed and recreated in post-production, the task is made much easier when the baseline effects of crunching, twisted metal are already there. Having those sounds in such well-recorded form likely gave the sound post team a great foundation on which to build.

It should also be noted that this was a second unit situation, with stunt people driving the cars and responsible for all the on-camera action. The principal actors weren't even on set at that time, and there wasn't any dialogue to be recorded, so not having that task gave the Sound Department time to do something useful. In this situation, one could question why the Sound Department was even needed on set that day; it would have been less expensive for the production not to have them there. A scene like this could have easily been shot MOS – early film-speak for shooting without sound, literally, "mit out sound," after an early German director's fractured English – see the Glossary (Chapter 6) for more.

In the end, it is essential to have a sound crew there in case the need suddenly arises for a line of dialogue or something unforeseen occurs. It's also helpful for the editorial team to have at least a guide track of the sound that happened that day; they can also hear the direction being shouted out, along with other audio cues. Again, the post-production crew was undoubtedly glad to have at least the option of having the "real" sound from the set to work from.

In the end, the goal of the sound professional, in both verité/documentary and narrative production situations, should be first to get the cleanest dialogue recording possible. Everything else is secondary to that cause and should only be pursued if it can be done without slowing down or compromising the production.

2.5. POST-PRODUCTION PERSONNEL AND HOW THE WORK GETS DONE

A lot of work and creativity go into creating a compelling soundtrack, and the creative choices that bring us there can seem nearly endless. At the same time, technology has evolved to better-integrated workflows that move the work along faster, more efficiently, and with higher-fidelity sound, providing greater detail and even more options in the sound build.

While technological advancements have been significant, an essential point in post-production and the sound design process is that the team is only at its creative best when all members can effectively work together. It's then that the collaboration between individuals of different expertise and their accumulated creative skills can all come together to create something transcendent. Then, they'll have a soundtrack far greater than just the sum of its parts.

The post-production sound team often waits to begin their work until the picture has locked, meaning there'll be no further edits; the timing won't change from now on. In a business where time is money, changes to the timeline can become time-consuming and, therefore, more expensive to accommodate

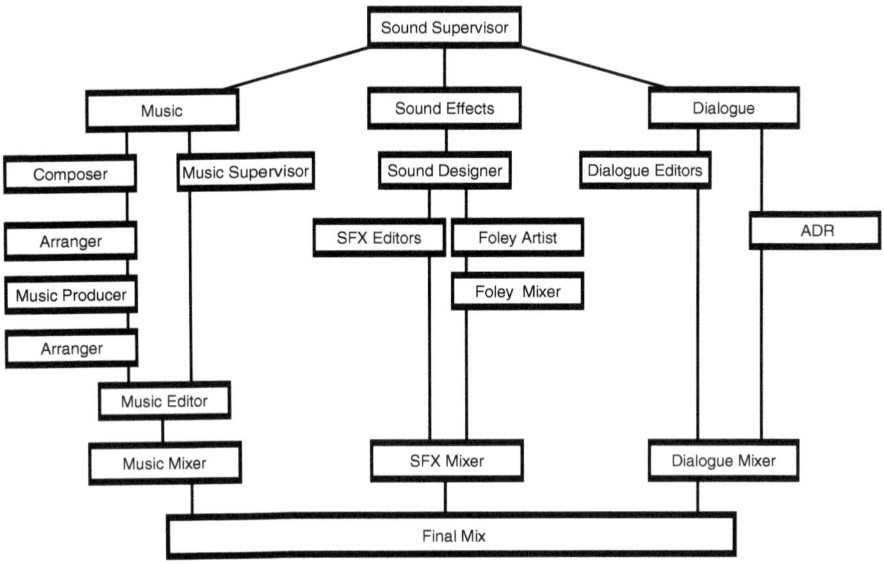

FIGURE 2.11 Audio post-production workflow

when they affect sound work that has already been done. The picture edit should ideally be "locked" before the critical sound work even begins, meaning only that the timing and timeline of the story have been set in place and won't be changing (much) from this point forward. However, mandates like this don't always hold, and picture changes often reveal themselves later in the post-production process than we might otherwise prefer.

That's not to say sound, editorial, and VFX can't all work concurrently, but communication between the different camps should be clear, and changes shouldn't happen without all sides knowing what's coming. Those changes can become problematic for the audio post team because they deal with so many tracks and timelines (see Figure 2.11).

All those audio streams need to be brought into sync with subsequent picture changes, which introduces a potential new source of compounding errors when all the teams and their tracks are considered. This is to say nothing of the time spent correcting the soundtrack and re-syncing it to the picture cut.

In some situations, a hybrid workflow could benefit the story. Mark Mangini has talked about how several scenes came together with the help of the editorial and VFX teams when he worked on *Dune* with director Denis Villeneuve, with the sound design team coming on far sooner in the production workflow than is typical – in this case, while the film was still in production. On that film, Mangini says, "We would build sound even before the score was created. Sound would then influence the edit, and we created a very symbiotic circle of sharing." The two teams – editorial and sound – would trade scenes back and forth, each taking the work the other had done and then adding on their work, improving it slightly in the process. Picture could be nipped and tucked, following the lead of sound at times, and sound would be tailored to better fit the edit that would come out of that process.

It has been a regimented process in the past, with picture lock nearly always preceding any sound work. Mangini and his team are working to make the process more collaborative by having all the post-production teams interact more as they build their story, most typically on the broader strokes of the overall sound design process. The back-and-forth better serves events and situations that are fantastical and have never been heard before, from distant planets to spaceships to giant sandworms; it could be argued that the VFX and editorial teams needed all the inputs available to help motivate these elements and bring them to life.

Other links in the chain, such as the dialogue edit, the Foley shoot, and incidental sound effects editing, just by their very nature of being more precise and usually synchronous processes, would all be better served coming later in the process, once the picture is truly locked and the director has signed off on it. This also comes down to economics, as SFX editors, actors being called upon for dialogue replacement, and Foley artists (and their stages) are all costly; it will exert some strain on the budget to have them all come back more than once.

Further refinement of other visual story elements will continue as the film story goes through the online process, with color correction, transitions, subtitles, and other special visual effects continuing to be refined and worked on. Like the Sound Department, the VFX and Color Departments prefer a locked cut so they're not doubling up on the work and adding unnecessary costs to the budget. At this stage of the process, picture timing changes can feel like "one step forward, two steps back" in that they often create a great deal more work for all the departments.

Visual storytelling is most effective when the group truly collaborates on the most creative aspects and the technological drudgery is handled. This happens within the sound team and in all the interactions across the entire post-production process. All the parts and teams must be moving in the same direction and toward the same goal at the same time. Creativity is essential, but to facilitate and nurture that process, so too are the organization and leadership, which will give the necessary space for that creativity to happen.

2.5.1. Sound Supervisor, Sound Designer

At the top of the post-production sound team is the *sound supervisor*, whose job is to keep the team on track, organized, and together, logistically and creatively. The job is a combination of administrative duties, with a lot of detailing and managing costs, organizing personnel, and, at the same time, bringing the director's creative vision to life as well. The *sound designer* might well be the same person, in the same position, working in the same supervisory role; it sometimes just comes down to the preference of the person doing the job. In some films, it could very well be two positions, with the sound supervisor handling the nuts and bolts of the work while the sound designer oversees the more creative aspects of the soundtrack's design and execution. Here, I use the two interchangeably, with the understanding that some who wear the title might take exception.

The title of sound supervisor harkens back to the early days of film, but over the years, the job has been redefined due to changes to the process, the technology, and the demands of modern storytelling. The role can vary somewhat from project to project, depending on the scope and needs of the story at hand. On some projects, the sound supervisor might come on board just as the post-production process is getting underway, while on others, the sound supervisor might be involved at a much earlier point in the process while the film is still in production. In nearly all cases, they're the first ones to join on the sound post side and the last to leave, overseeing all the mechanics and administration of the process through the final deliverables while realizing the creative vision along the way.

As the journey through the post-production process begins, the sound supervisor will need to assess the story's needs from a personnel standpoint and then "crew up" accordingly. They will need to organize and hire a team of editors, mixers, Foley artists, and others whom they feel are best suited for creating the soundtrack for the visual story.

Unfortunately, a sound supervisor's early involvement is too often seen as a budgetary luxury, serving as one of many examples of how art and commerce intersect and, sometimes, struggle for hegemony over each other. On the face of it, producers might wonder why this post-production sound person should be involved at all during the production phase, but the reality is that these early interactions can be a cost-effective strategy. This early involvement can head off some issues before they become problems while also helping to streamline the storytelling process. A sound supervisor's early input can have benefits that might otherwise only reveal themselves in a kind of creative hindsight, but this early involvement works best when it's in service to the story.

"I've been advocating for years on the value of starting sound as early as the screenwriting stage," says Mark Mangini. "I've been engaged to help at that point on a number of films with great success."

In the early days of film sound, the sound supervisor played a more administrative, organizational role, in large part because the process of film sound itself was a series of tedious, manual tasks that were time- and labor-intensive due to the mechanics of handling, cutting, and splicing the film itself. Sounds had to be physically spliced into the film or mag timeline, which was a lengthy and somewhat complicated physical process, and only after that process could a track be previewed as it was played from a machine, or "dubber," in sync with the picture, and then incorporated into the greater sound build. The process took a great deal of time, especially when compared to current methods and processes, and crews sometimes numbered in the dozens, with all those people answering to the sound supervisor, with the budget dictating how big the sound editing team could be to get the soundtrack into shape for the final mix.

The role of sound supervisor has expanded in modern times, with the position assuming more creative aspects of the process, but still in addition to the crew oversight and administrative details they had previously. Meanwhile, technology has allowed for smaller, more efficient crews, and these crews have more autonomy than their early-days counterparts would have had. In all scenarios, the sound supervisor serves as the main point of contact between the director and their creative team, all working to bring the director's vision to life.

Once the film is locked and in the hands of the sound supervisor and their team, many editors with different specialties swing into action. Dialogue, effects, and music teams can begin their work in earnest once the story has locked. They're all in a race to the finish, trying to get all the moving parts into place while working together and arriving on the mixing stage at the same time. There is an order to the process, with specific tasks needing to happen earlier in this progression before other jobs can even begin.

2.5.2. Dialogue Editing

Dialogue is perhaps the most important of the processes in most ways while also being the least understood. We rely on dialogue (and voiceover) as a central part of the narrative to help tell the story. We humans are deeply attuned to human speech; it's been with us since birth, and it has also been key to the survival of our species and our very civilization, so we all are very discerning as to how realistic speech and voice material should sound. As a result, we're also very good at recognizing low-quality dialogue or when the voice tracks have qualities that are just "off," somehow. In the end, if our production sound people did their job well, we will have clear, clean, and completely usable dialogue to work with, save for some cleanup and editing details.

The Dialogue Editors begin their work immediately after the picture is locked. Years of experience will help them discern what dialogue will work and what lines will need further work, but as they go through the dialogue, their mission is threefold:

- First, remove all non-dialogue sound and noise from the track, at least to the extent possible. If any non-dialogue sound on the production track is at all usable as part of the sound effects build, it will be pulled from the production tracks for the SFX crew to work with and incorporate into their sound builds, but the dialogue editors will work to isolate the dialogue from the rest of the production audio.

  As part of their editing process, the Dialogue Editor will also utilize a variety of processors and plug-ins to clean up the dialogue tracks, such as denoising, undistorting, removing other stray and unwanted sounds, and adjusting edit points, so that what ultimately remains is the cleanest dialogue possible.
- After all that work, we may find that some of the dialogue may not meet the quality levels we would like to see – or hear. In those cases, lines will be tagged for replacement in future *ADR* (Automated Dialogue Replacement) sessions.

  ADR is a process that is sometimes fraught, and directors from Robert Altman to Christopher Nolan have resisted or even outright refused to take part; they see it as compromising the original performance the actor gave on set. There is some validity to that claim, as it can be challenging for an actor to get back into the right headspace and become that character again, sometimes weeks or even months removed from when the original scene was shot. They will be re-performing those lines almost in a vacuum, with no interplay or interaction with other actors to help bring their performance back to that original energy level and emotion.

  The dialogue editor will examine other takes of the same scene to minimize the need for potential ADR sessions. This can save some time and effort for both the actor and the ADR team, as the substituted dialogue would have the same qualities – energy level, ambient sound, and so forth – as the original.
- The dialogue editor will then balance the production dialogue, smoothing edits left by the picture editors and adjusting levels for consistency while applying some signal processing to help intelligibility and perspective. Once ADR is complete, they will incorporate their edits into the overall dialogue build, working to make those new tracks work with the rest of the dialogue as seamlessly as possible.

This final step sets up the dialogue for the rest of the track build and is the last stop before beginning the predubbing process. Dialogue editors may publish their works in progress and share them with the rest of the sound post team as they go, just to ensure everyone is on the same page and has the latest version of their work.

2.5.3. Foley

Foley is the process of adding non-dialogue sound elements to a track to help make it more credible. The kinds of sounds the Foley team will work on are the more nuanced and generally human-generated sounds that don't lend themselves as well to "canned" sound effects – that is, sound effects that traditionally come from sound effects libraries. Sounds like footsteps, body movements, and other physical interactions our actors on screen might have with their environment and items they handle are more easily performed and recorded than trying to shoehorn library sounds into place; the nuances the Foley team can bring will make them more believable.

The process is named after Jack Foley, a member of the first sound effects department at Universal Studios. Mr. Foley found it more straightforward to perform sounds in synchronization with the picture than mechanically editing pre-recorded sound effects into place, which helped enhance the action on screen. This was also at a time when the Sound Department had more authority on set, taking the mission of capturing clean dialogue, and *only* clean dialogue, very seriously. Production audio sounded very sparse on its own, however, and there was little sound presence otherwise. Consequently, there was little to compel the viewer to believe that what they saw was real.

This concept of adding sound to picture was a new idea, and it became a central selling point for the motion picture studios, as they were all looking for an edge over the competition. Mr. Foley's work evolved into more extensive sound effects editing in later years as the motion picture studios built up their own sound effects libraries, with these banks of sounds becoming valued resources for the studios, each one of them competing to gain any edge over their competition they could with their proprietary in-house sounds.

One of the most common examples of natural, incidental sound is footsteps, and the term Foley has become synonymous with footsteps to many, but Mr. Foley and his team would take on a variety of sound effects in their work. Any movements or actions of the actors on screen that would typically make a sound were given the "Foley treatment" and then added to the soundtrack to help make it more realistic and believable. "Send it down to Foley" became the catchphrase when a scene needed extra sound support, and in very short order, Mr. Foley's name became a noun, a verb, and an adjective. We now have the process known as *Foley*, where we Foley sound effects onto a Foley track as part of our Foley work.

The first step that the modern Foley crew will embark on is to *cue* the film in a process sometimes referred to as *spotting*. "I'll cue the whole film from top to bottom," says a Foley artist and mixer, Mark McBryde. A team member, often the Foley mixer, will watch the film, painstakingly marking all potential Foley cues for the session to come, which they will coordinate and record with the

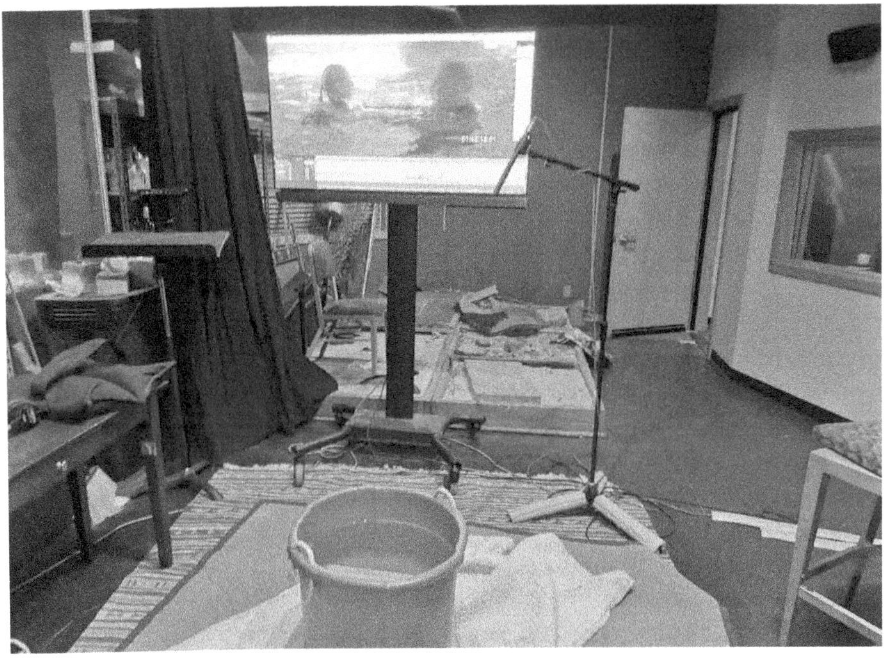

FIGURE 2.12 A typical Foley setup

Source: Courtesy Mark McBryde.

Foley artist. "We'll watch the whole film down, almost frame by frame, so I'll know it when we go into session," says Ryan Maguire, Foley mixer and editor. The planning stage is key to good workflow and keeping the Foley session on track. "So much time savings goes into that," says Maguire. "We can save a day or more on the stage if we plan it out properly from the cueing session."

That Foley "cue," which is a collection of multiple cues, will then be taken into the session, or Foley shoot, where the Foley artist will draw on an array of sound props, devices, and surfaces available to help in the creation of original, bespoke sound effects that exactly match the action being seen on screen.

The space where the Foley session occurs is the Foley stage (Figure 2.12), a purpose-built recording space that differs from traditional recording studio spaces in a couple of ways. The acoustic treatments will be similar, isolating the space from the outside world while controlling reflections within, but with some notable additions. There will be several different floor surfaces and plenty of storage and organizers for the many Foley props that are helpful to have on hand.

"First, I'll record a cloth track, then feet, then props and hands," says McBryde. "Footsteps will be recorded for everybody from top to bottom; then I'll record the hands throughout the whole movie. And by that, I mean everything they touch, move, or anytime someone puts their hands down, touches their skin – we cover every hand motion, from top to bottom."

The general rule of thumb for sound effects work is: "if it moves, it makes a sound." While sourcing individual sounds might be up for grabs between the Foley and SFX editing teams, some sounds will be better served by sound effects libraries. Sounds such as gunshots, explosions, and breaking glass are generally messy, destructive, possibly expensive, and even dangerous in enclosed spaces and are best sourced from sound effects libraries. Foley is better suited to the more interactive sounds that are caused by the actors, such as the aforementioned footsteps, props, and hand movements, but also cloth and clothing rustles and specific actions, such as fight scenes, sitting and standing motions ("sits" and "ups"), falls (both bodies and objects), and so on.

Choosing Foley sound props has different criteria than those a set decorator might use to choose the props for a visual element in a scene. In the world of Foley creation, our props only need to make the right sound, and we don't care what they look like. In fact, they might not have anything in common with the objects we see in the scene.

Foley artists will go to great lengths to ensure the sound is right for what we see, even if the object being used might bear little resemblance to the object seen on screen.

A great example of this would be gun-handling sounds. Foley artists are not concerned with capturing the literal sound of the gun being fired – remember those "messy and dangerous" properties – but they do provide all the other sounds that are a part of that. Guns and gun handling are almost a sound cliche because we, the viewing public, have been conditioned to believe that a gun makes all sorts of metallic rattles and clicks when it's being moved about. The reality is precisely the opposite: a gun in good repair in the real world makes little sound until it is loaded and fired.

To achieve the metallic gun-handling sounds we believe a gun *should* be making, many Foley artists may provide gun-handling sounds through any number of everyday objects – one common trick is to use a standard, industrial-grade staple gun. That staple gun has a lot of loose parts, so it makes a whole host of satisfying metallic rattles and clicks that our eye/ear/brain combination interprets as the perfect collection of sounds for the action and motion we see on screen. For the uninitiated, a typical Foley stage is a cluttered, even chaotic place, with an extensive collection of seemingly unconnected props lying around, with many of those objects' purpose impossible to guess at first glance.

"Once we've finished the show, it's off to the Foley editor and then back to me," says Ryan Maguire, "and then it's off to the dub stage." It will then be added to the overall track build and balanced to fit in and around the other elements as part of the pre-mixing process. There's no better indication of a Foley track well done than having no one notice the work in the final mix, something all Foley artists would agree is precisely the point. They're trying to create a world within the film that the viewer buys into, completely and without question. Good Foley furthers the illusion and immerses the viewer in the world of the story we're trying to tell.

Another requirement that necessitates the Foley department's best work is the deliverable for international distribution, called the *FFE*, or Fully Filled Effects track. Some incidental sounds might be on the production track, where they were inadvertently captured, but not typically in any usable form. The sound crew on location is most concerned with getting clean dialogue in production – everything else is incidental – so when that production dialogue track is removed, as will be happening on any international version, it will fall to the Foley crew to cue and capture all the sounds that help make a scene feel like reality. "If it's in the original domestic version, then it needs to be in the M&E stem," says Mark McBryde. The M&E track is the Music and Effects version of the track, without any dialogue. The dialogue track can then be replaced, or *re-dubbed*, into the language to suit that market.

At the end of post-production, a film will undergo an evaluation process as a level of quality control to ensure it meets all technical criteria and aesthetic standards. Foley is essential to this process because it provides a sort of sonic credibility by filling in all the spots that weren't covered in production. The international version of a film needs to sound identical to the original version of the film, minus the dialogue, with all character-created sounds intact to sell the idea that it all happened at the same time and in the same place.

And so it goes, this journey of the Foley artist, in service to both technical requirements and creative desires. All in pursuit of the perfect sound for an on-screen action, using an array of outside-the-box thinking and unconventional sound props to get us, the viewing public, to a place of believability.

2.5.4. SFX Editors

As with the rest of the audio post-production workflow, sound effects editing was originally an analog process on a physical medium that could only be executed linearly. We could only access points in a film reel in real time, or perhaps double time; it could take a few minutes just to re-cue a reel and rewind it to the beginning. The track had to be physically spliced together with timed leader between sounds so it could play in sync with the picture or even be playable at all. If something needed to be corrected, such as the wrong sound being used or a sound landing in the wrong spot because of some mistimed length of leader, the mix session would have to be stopped to address the issue. The whole reel would have to be taken down, re-edited, and timed again before being strung back up and replayed.

Today, we have any number of platforms that will record, edit, and place sounds in absolute synchronization with each other and with picture in the same operation. Options can be tried repeatedly, making it feel like there are sometimes infinite possibilities for that finished soundtrack. However, this concept of endless options can sometimes create a logjam of creative ideas.

Whatever the mode, analog or digital, it's good to approach any sound design with a solid plan in place. Indeed, we want to allow some time to experiment and allow for some serendipitous things to happen, but at the same time, there needs to be a focus on exactly what the plan is to bring all the sound elements together. We can begin by categorizing and delegating tasks, with the division of labor becoming ever more granular and specialized as we drill down through the layers of sound we will need.

Foley comprises a significant part of the sound effects track. When the final Foley session has been completed, those tracks will be incorporated into the greater sound effects build by the sound effects team. The sound effects editors might also use those Foley effects as a base for their sound effects build, layering them with other sound effects elements.

The SFX team will sometimes work on effects groups by specialty or expertise, with some of those assignments going to effects editors with greater experience and knowledge in certain sound areas. For example, an editor might be known for their horse sound effects editing, so they may get many calls for work where horse sounds are needed. On the one hand, there is a kind of job security to having a skill like that, but on the other, this business tends to lead to typecasting in the behind-the-scenes positions as well. It can prove challenging for an editor who might want to branch out and not necessarily do just horse sounds all the time.

Beyond the Foley work, sound effects editing tasks might get further broken down into hard effects, soft effects, and ambient sounds. Note that these are mainly organizational tools, for the most part, used to help instill order in what could otherwise be a chaotic sound build. From this point, further subdivisions and definitions can (and will) become even more specific, with all imaginable categories; a superhero blockbuster might have over-the-top fight sounds, as one example, whereas a movie about cars and racing will have several subcategories of sounds to help define the vehicles, from engine sounds, mechanical and other related sounds, to tires, doors, wind noise, and so forth.

The accurate portrayal on screen can take a lot of work; for example, cars are complicated machines and a great example of how the sound effects team might need to take care in their recording sessions to get the appropriate sounds. That would mean acquiring the actual vehicles seen on screen (or soundalikes) and putting them through the motions seen on screen to create all the sounds that those cars need to make in the context of the film. The suite of sounds for the car would include recording it from every possible different angle. For example, a car sounds different from the front than it does from the rear, and the interior has a wholly different quality than the exterior angles. There are also all the possible modes of operation, as the engine accelerates, decelerates, idles, and drives at speed. At the same time, all the mechanical aspects need to be covered, including the doors, hood, trunk opening and closing, the sound of switches, the shifter, pedals, door handles, and other processes, like wipers running and windows opening or closing. This collection of sounds is called a sound workup and obviously can be extensive. Many of us are used to at least being around cars, so we can readily identify when a car sound is correct and believable or if it somehow misses its mark.

The editors who deal with ambient sounds will need to work closely with the dialogue team to get the interplay between voices, room tones, and environmental elements to be as natural and believable as possible. Ambient sounds, also sometimes called ambo or atmos (not to be confused with Dolby's immersive sound projection technology), form the underpinnings of any given scene, as they help establish its geography, timing, and sometimes even mood and emotional context.

Lastly, we have designed elements that can run the gamut from science fiction to spooky to otherworldly sounds no one has ever heard. At times, these types of sounds can exist between the world of sound effects and the world of the music department, as certain elements can have tonal or even musical qualities. It's always best to have some synergy between the music and sound effects departments, if possible, so ideas (and musical keys) can be bounced off each other and the results of both work together.

Ultimately, every sound is another opportunity to help tell the story. The quality and character of each sound and finding the emotional content in an otherwise mundane sound effect can influence the feel of a scene and help a visual story get to a place where the overall impact is visceral, emotional, and authentic.

2.5.5. Predubbing

The idea of predubbing is to lighten the load for the final mixdown session by making many smaller mix decisions along the way, creating and improving the mix in increments and steps. This is an evolving process, where the sound design and mix happen more organically as the sound team can hear the interaction of the sound elements along the way. Attempting to balance a bunch of seemingly disparate sounds together at one time on a large format console shouldn't happen; all those sounds should grow together, and all have a place in the final mix print. Through the process, the placement of each sound with the rest of the sound build will initiate a new round of reassessment and adjustment. Similar sounds will be worked out relative to each other in groups at first and then against the other groups of sounds as the refinements become less granular and more holistic.

For example, the soundtrack for a modern film is often comprised of dozens of elements or tracks of sounds, and many of those sounds sometimes lend very fine detail to the overall sound effect or element at hand. Sounds like the creak of a door or a bit of fabric rustle as a character moves through the scene can be challenging to keep track of individually, particularly when so many of these fine details must be dealt with simultaneously.

"You couldn't have the kinds of soundtracks we create today with hundreds of tracks without predubbing," says Mark Mangini. The level of detail is too great and simply too much to manage at once.

It starts in the sound effects edit, where sound editors will do their best to balance their sounds and set levels as they are placed to picture. This is one of the best audio track management tools the sound editor has, ensuring that others downstream in the process don't have to guess that editor's intentions; the initial mix level is already baked in. Documentation certainly helps, but there are limits to how helpful pages of notes might be to either end of the process – producing them is time-consuming, as is reading and implementing them. Besides, that editor likely will have moved on to their next project by the time the final mix rolls around, so they might not even be available to ask, making it even more critical that all tracks be as intuitively laid out and presented as possible.

The standard we all try to follow is that the tracks we work on present at a nominal reference point – best known by its analog identity, unity gain, seen in Figure 2.13 as the 0 mark on a typical fader.

With this as our starting point, we already know the original editor's intentions, so by presenting all the tracks in this way, we have the beginnings of a coherent mix. From these beginnings, we can begin to see how the term predubbing came into being. At its simplest, it's a process for making mix decisions in smaller, more manageable chunks and pieces.

It might help to have a little historical context first.

In the earlier days of visuals on film, the media that carried the audio was also based on celluloid film, except it would be coated with iron oxide, much like magnetic analog recording tape. The editing process existed entirely in the physical domain and was destructive – a razor blade was used to cut the magnetic film (or simply "mag") to make the edit, splicing the ends back together with very thin adhesive tape. Edits could be undone, but restoring a piece to its pre-edited state and returning to a previous version of a track could be time-consuming, as any piece of audio might be misplaced or lost among other scraps of media on the cutting room floor.

The editing systems that facilitated this process were unwieldy and expensive, as were the machines used to play each finished strip of mag film,

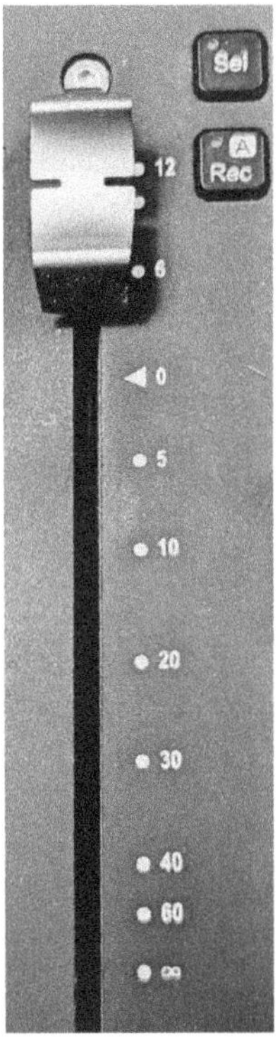

FIGURE 2.13 The standard audio fader

or *track*, on the mixing stage (see Figure 2.14). Audio post-production was a costly proposition, and having more tracks available required adding even more of these large, expensive machines for the playback. To get around this, predubbing was perfected, where multiple mag tracks would be mixed and recorded – or dubbed – onto another deck. In theory, this process could go on for as many tracks as needed and provide the mixer with a richer, more complex sound build than a limited number of machines would have been capable of producing.

Innovations and refinements occurred along the way as the technologies around film sound evolved, but it remained a process with physical limitations. This hardware required space and resources, with initial cost, maintenance, upkeep, electricity, operation, and cooling all being significant factors. Film dubbers were also extremely noisy, so they needed to be relegated to sonically isolated spaces so as not to interfere with the mixing process; they were typically housed in a machine room with a separate air conditioning system and kept as far from the mixing environment as possible. They were analog audio

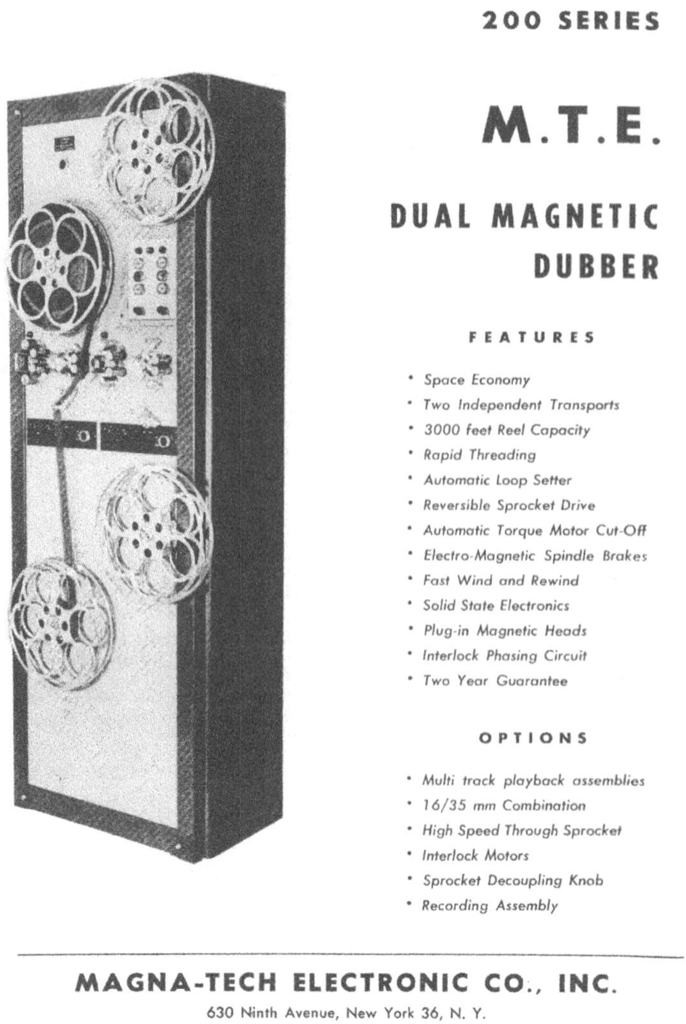

FIGURE 2.14 Magnatech mag dubber – one machine represents a track or two. The more tracks needed, the more physical machines.

Source: Magnatech Corp.

devices, so there was also a significant amount of noise inherent in the recording process itself, and that noise would be compounded with each additional, re-recorded layer. Steps and techniques could be used to minimize that noise buildup, but it would always be a factor and would only get worse with each generation.

The advent of digital technologies eliminated the analog noise issue, but the early versions of that technology were even more expensive, as this new technology was still evolving. They often went through bespoke, typically hand-built analog mixing surfaces, adding another layer of expense to the process.

Historically, the film world has been slow to adopt new technologies, due in no small part to an inherent bias, a kind of purist resistance to new modes of doing things – the industry as a whole is adverse to the risk potential of unproven tech. Downtime means cost, and with potential fixes and workarounds being less proven, there was a definite need for the tech to prove itself before widespread adoption could happen in the broader community. Many might bemoan the sometimes glacial pace of change, but the upside is that the tech that does make it through these challenges will have been carefully vetted, tested, and, ultimately, the most reliable work path.

Technology marches on, and the industry has gotten to a much better place as a result. The tech found in a mix room or stage is something that would have been almost unimaginable even a short time ago, and the means to work with many (so many!) audio tracks or streams simultaneously has reached a point where, in many cases, it even exceeds what is needed.

At whatever point in the process mix decisions may be made, from editorial to actively mixing, the ability to execute all those options and experiment has been the most significant innovation of all.

2.5.6. Mixing (or Re-recording or Dubbing)

It comes down to a question of: how prepped do we like to be?

When working in the analog domain, such as tape or mag, sound mixers developed workarounds, such as working on some signal processing while playing in reverse, a frequent practice of the times. It wasn't necessary to understand the words being said in reverse, but that voice is still a voice, and processing was based more on the frequency content and elements of the voice than on the words being spoken; the direction of play didn't matter.

In the modern-day digital realm, the workflow options are nearly unlimited. We still need to listen in real-time, but nonlinear access to the material means no rewinding, making all points in the film's timeline instantly accessible. Adjustments can be made and refined; they're all stored in the automation data that can be revised and modified nearly endlessly. Track counts are no longer limited by physical media (and the machines needed to play the media on); digital workstation tracks can number into the hundreds if needed. Signal processing isn't limited to a few specialized (and costly) pieces of hardware; signal processing plug-ins can take us beyond that limitation without the expense or physical space requirement.

2.6. TOOLS OF THE TRADE

2.6.1. Workstations

There are many digital audio workstations and platforms to choose from, with the decision falling to the user as to which platform works best for them, both technically and ergonomically. That said, there is a certain amount of compatibility required in all sound for visual work, depending on where the workflow is headed. That, in turn, will narrow the selections somewhat, sometimes down to the workstation itself.

Pro Tools is the current industry standard, and by a substantial margin. That could change, and another workstation might find dominance at some point. There may even be other workstations that can do the job for a given application just as well. It comes down to compatibility with the person operating it and the rest of the project's workflow. There's also the question of compatibility with the rest of the world – mainly whether a system's files can be incorporated into a larger project or session file and if the platform is compatible.

Many professional workflows incorporate AVID technologies, from editorial through sound post-production, and the easiest way to do work in that flow is to use the same platform. The work of editing sound can certainly happen on a different platform, with files converted to the session master format. In the end, all elements need to make the journey through that translation process – which can be challenging when working in the ecosystem of a different platform.

Let's say an editor is working on some sound effects for a larger project that will be mixed on a large mix stage later. The sound supervisor would likely stipulate that the work deliverable be in a particular format, with tracks laid out in a specific way and other technical specifications. To deliver that project properly for the master mix session, the project data need to be translated into that form as seamlessly as possible. The audio files in their edited forms are the most critical element, but any plug-ins or signal processing that happened along the way must also make the conversion journey.

Many sound supers might stipulate that all files be transferred preprocessed and at mixed level so the re-recording mixer on the other end can import those tracks into the final track assembly, and the intended mix will be heard immediately without any adjustments. Preprocessing is the ideal, as there are so many plug-ins available – some can be somewhat obscure, and it is unrealistic to expect every system to have every plug-in ever created and immediately available.

The subcontracted editor would discuss the session file deliverable with the sound supervisor in advance. Is a specific session file format required? The most common currently is Pro Tools, but it's critical to the process to establish early on if it will be something different. Likely, they would require a "flat" session without any active plug-ins, with any signal processing rendered as new audio files and no active mixing data. In case of a workstation incompatibility, will they accept other file formats, such as PTX, AAF or OMF? What about the track layout? Do they have a specific session template or format they'd like you to follow?

If the final mix is happening on a mix stage or needs to be incorporated into a more extensive track build, following a template further helps standardize the workflow so everyone knows what to expect and where to find specific elements. The final track layout and workstation format choices are up to the sound supervisor and the mixers; they'll go with the most comfortable, affordable, and technically capable platform available for the task.

There are many platforms to choose from; one should evaluate the demands of the work being done and then find the platform that best suits the needs of the job at hand.

First, consider track count and the physical capabilities of the platform. A feature film can undoubtedly eat up dozens, even hundreds of tracks once everything is broken down to the sound macro groups of Dialogue, Music, and Effects (DME). Dialogue can occupy 12–24 tracks by itself – even more, when all the production dialogue, alternate lines, and ADR are accounted for. Then, further organize those tracks by breaking them out by character and location. A simple layout of those tracks might see our main characters each on their own track, with any ADR tracks adjacent and further broken out by scene and location. One basic mode of track organization would be moving them into two main groups, one for interior tracks and another for exterior tracks. There could be additional lines, as anything that is discernibly language-specific will need to be included in the dialogue track, such as crowd voices or singing. Foley elements and sound effects can easily consume many more tracks, with ambience taking up another swath of track space.

We could see another two or more dozen tracks there, depending on how the music is delivered. Some music cues may come in as stereo mixes, but composers often provide individual music stems for the added flexibility of being able to edit or even move music elements within their work if needed. For instance, the melody track of a piece might conflict with dialogue in a scene – the first impulse would be to lower the entire music track, but that would reduce the impact of the music cue overall. It's better to minimize the conflicting elements of the track – in this case, the dialogue and melody seem to be in conflict – so we can maintain the impact of the music cue, while lowering the melody and getting it out of the way.

With all these inputs, we also need to consider the system outputs available. The mix must be accommodated in all its forms, from mono to stereo to 5.1, 7.1, ATMOS, and beyond, with each format taking up ever greater numbers of outputs. Often, we can mix in more than one format at once; it's up to the limitations of the system and the processing that converts from one mix format to another. Some plug-ins will do this, but in the end, it's only an algorithm and may require human intervention.

"We require a hands-on version of the 7.1 mix, and a hands-on version of the 5.1, and so on," says Mark Mangini. "There are algorithms available to do those mixdowns from ATMOS to 7.1 to 5.1 and stereo, but we don't like the sound of them. There should be human creative intervention for each one of those."

Any track building will benefit significantly from organization and workflow standardization. Keeping the elements organized and grouped like this (see Figure 2.15) will help facilitate the most complicated of track edits and sound design, accomplishing two goals. First, it will set us up for deliverables by keeping all the sounds organized by family – remember the DME (Dialogue, Music, and Effects) stems that typically need to be delivered – which will further speed up the deliverables process (see Figure 2.16). Second, it will provide easy access to the material, and that streamlined workflow will allow everyone to quickly and easily locate and adjust any sound elements within the track build.

Depending on the scope and scale of a project, the system may or may not be able to accommodate the means to the project's end goal. All platforms have their limitations in track count limits, routing options, and signal processing power. Some systems can be chained together to create a more powerful system; others require more elaborate workarounds, making the predubbing process essential to get past those workstation limitations.

Here are some important considerations to account for when assessing potential digital audio workstation platforms:

- How proprietary is the system? Some workstations are purpose-built systems running on their own bespoke hardware. While these are excellent machines, they tend to be expensive and exist in their own ecosystem, which can also be costly. They may not accommodate third-party peripherals or plug-ins.
- If it's not a proprietary system, what computer platform does the software require to run? Consider a more robust system than just the minimum processing and memory required. Those are nominal requirements and it's virtually guaranteed that the system will bog down under heavier demand; you may need to upgrade to a more powerful computer to run it. There's the perennial question of Mac or PC, and while we all have our preferences, the system may have a longer history on one platform over the other. The platform software is undoubtedly a significant expense, but the greater cost is always going to be acquiring the computing power necessary to make it run.

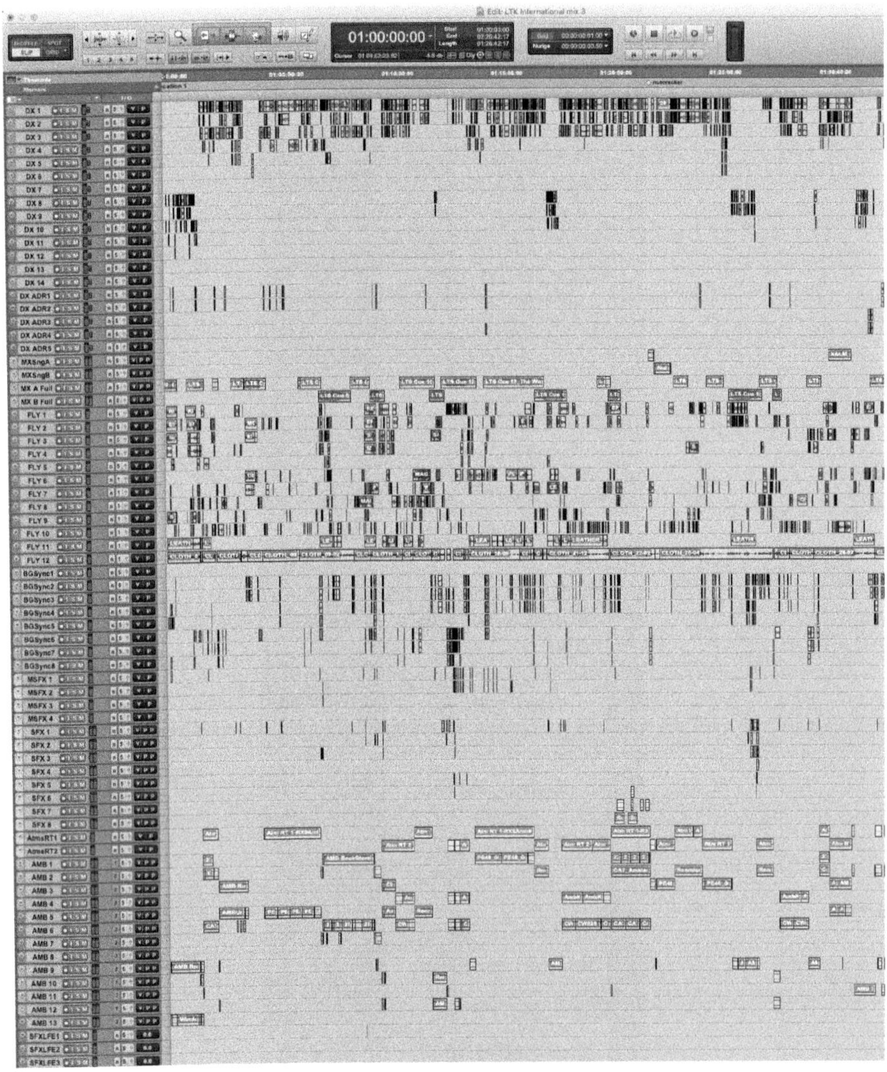

FIGURE 2.15 A typical Pro Tools track build

- It's important to consider the workstation's architecture and whether it's open enough to run third-party applications and plug-ins. Many plug-ins can run on a variety of systems in DSP, AAX, or VST modes, but not all plug-ins are created equal, and not all systems support all formats.
- Assess the number of outputs needed. Large-scale narrative work that plays in a large-format medium, such as theatrical, broadcast, and streaming outlets, will require delivery in surround, stereo, or multichannel formats.

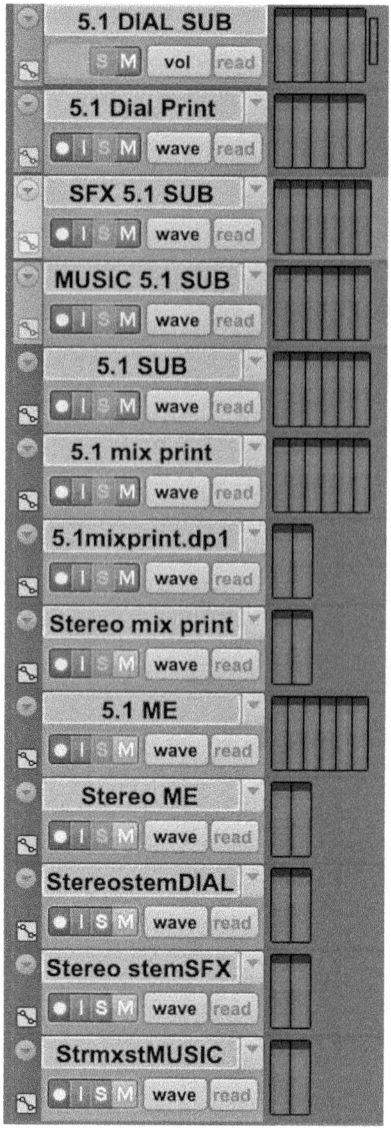

FIGURE 2.16 Pro Tools outputs as part of a mix template. The separation of stems at this stage makes outputting them a simple, one-pass process

The system must be capable of running many outputs simultaneously and in real time without glitches or interruption.

- This also gets to what the ultimate workflow might look like in terms of compatibility and, later, delivery. Does it need to integrate into a larger post-production facility with its in-house ecosystem established? If the primary workflow and delivery are to an AVID house, then that might point toward working in Pro Tools so that there's instant compatibility with those existing workflows. When collaborating with editors who work mainly in Adobe Premiere, installing Adobe Audition is the right move.

- Lastly, and certainly not the least consideration, is the overall cost of the platform. Again, the computing side of the equation represents the largest outlay, but the accessories, plug-ins, system interfaces, and the platform itself can also begin to add up quickly. Many platform software packages are migrating to a subscription-based model, which will help keep the upgrade schedule up to date. This will represent a regular expense over time that will need to be managed, so the peripherals required by the work being done will also add to the bottom line. Some basic signal processors will be required, as well as some more advanced packages, so it's important to ascertain what is included within the system software and what will need to be purchased separately.

2.6.2. The Foley Stage

The Foley stage (Figure 2.17) is an interesting place, designed for performance, acoustics, and props storage, all at the same time. The typical stage will have cabinets and bins for keeping all those props close by and different floor surfaces to perform footsteps and other actions. Many Foley stages will have interchangeable, moveable surfaces that include different kinds of wood, concrete, tile, and many others, along with Foley "pits" that can be filled with various materials, often including sand, gravel, and crushed rock. Some Foley stages have "wet" capabilities, either with tanks or with other pits and surfaces, meaning they can be used for water-based effects, such as splashes, walking in wet conditions, and more.

Every Foley artist will bring their own highly individualized collection of props, or "kit" (see Figure 2.18), to the session, also known as the *Foley shoot*. Some artists can be secretive about their kit, whereas others are open about what they do.

2.6.3. The VO Booth and ADR Stage

Sometimes, a booth can do double duty, serving as a voiceover and a dialogue replacement recording space. Sometimes they are different spaces, in different studios, specialized for a specific type of work, with slightly different microphones and setups. The common thread here is that all recording spaces

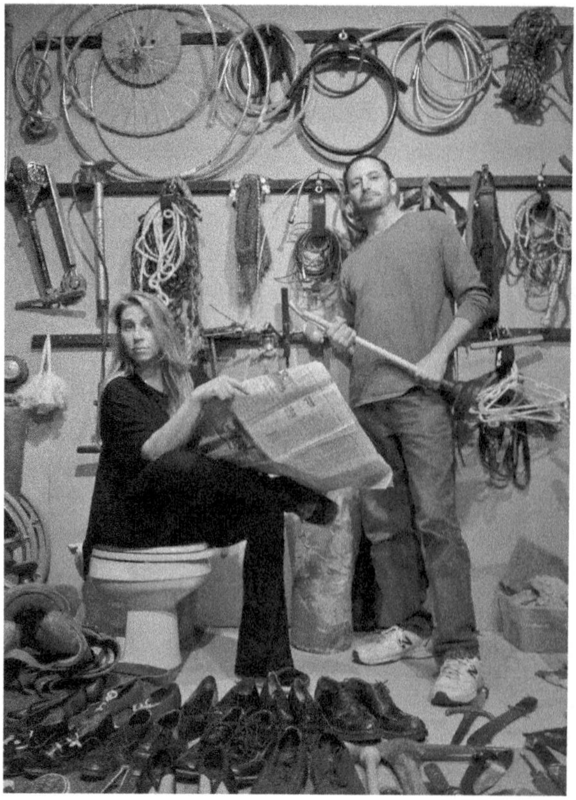

FIGURE 2.17 Tara Blume and Ryan Maguire, seen here hamming it up on their Foley stage

Source: Stephen Borasch photo.

are as sonically neutral as possible so as not to impose room acoustics onto the recorded dialogue or narration. In the case of dialogue, the mixer can later create the sound signature, adding reverb and blending different reflections to match better with the production audio.

The mixer may need to create a sense of space that wasn't present in the production track at all, but is necessary to the narrative. Perhaps the scene unfolds in a cave, with all the attendant reverberations, but the set on the isolated soundstage didn't have any sound reflections. This is the ideal scenario, where the dialogue team has the ultimate control over the reverberation and reflections, also helping to strengthen the scene's premise.

By comparison, a narration booth only needs one good, studio-grade microphone. In contrast, an ADR booth will have some options designed to better match the sound signature of the production mixer's equipment. The

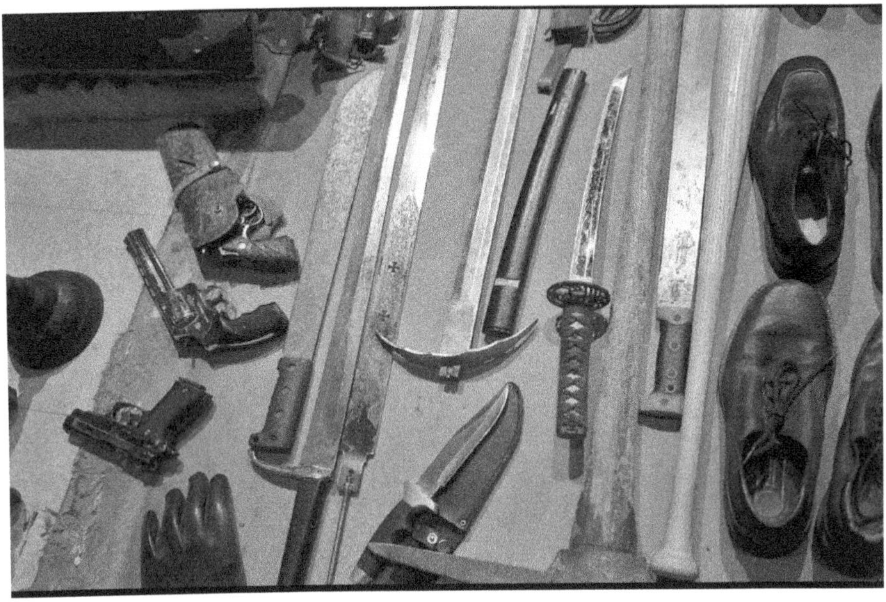

FIGURE 2.18 Some of Tara Blume's Foley props

Source: Stephen Borasch photo.

production team likely noted the microphone equipment they used on set so the ADR mixer could set up the same or similar-sounding microphones in the booth.

The ADR booth will also have a large-format video display so the actor can see – and match – their original performance and lip sync (Figure 2.19). There will also be a cueing system in place, which can be any combination of streamers superimposed visually on the video, audible beeps for a countdown to the cue, or cue lights in front of the actor.

Several software packages are designed to help with the ADR session recording process by providing some mode of countdown, executing the record on the workstation, and keeping track of the takes as they are recorded. Third-party plug-ins like ADR Master insert the text line onto the screen so the actor doesn't have to look down and away from the microphone while they do their line.

The trick with any ADR session is to achieve a performance that rises to a level that works with the performance seen on set at the time. This goes beyond the technology choices we might make for the session; it's entirely up to the actor to perform in a way that matches what they did at the time on set.

Working against this ideal is the tendency in a recording studio booth to talk quieter – it is a soundproof room with little to no sonic input from the outside world. The challenge is for the actor to "talk up" and perform their lines in a way

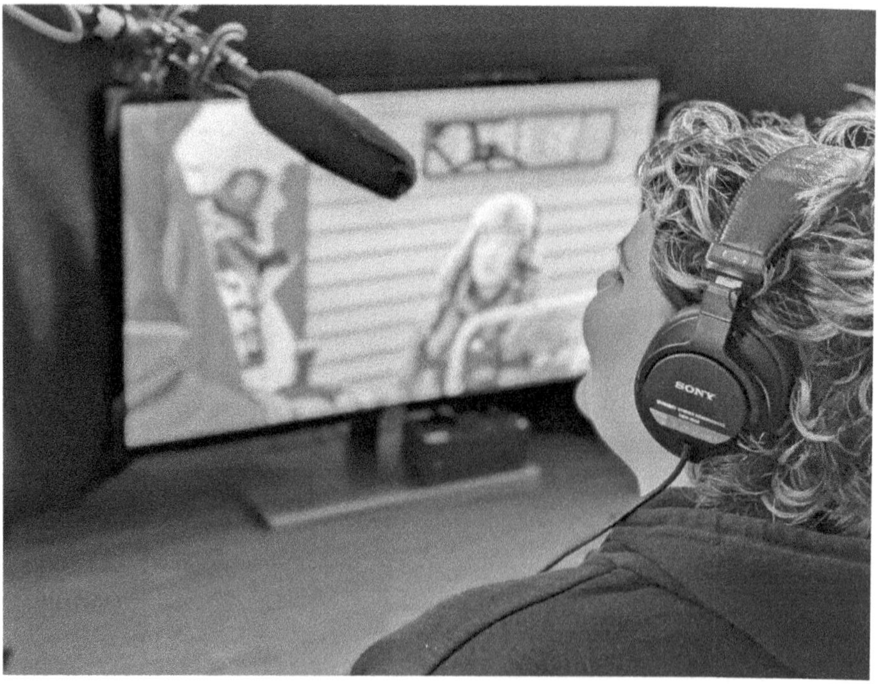

FIGURE 2.19 An actor stands in front of a screen in a recording booth, ready to re-record her lines

Source: Permission H9 Films and Moody Street Pictures. Photo courtesy of the author.

that will match their projection level on set. Sometimes, they might need a little help with their process, and feeding ambient sound from the production track or other sources into their headphones can help with that. The natural tendency when wearing headphones is to talk louder, and feeding background sound to them at the approximate level they might have experienced on set, along with the other actors' voices, can counter the tendency to be quieter in the booth and help them better dial in their projection.

2.6.4. Dubbing Stages

Also called mixing stages, this is the place where it all comes together (see Figure 2.20).

The final mix is where it all comes together, but on a larger feature, the mix might happen through the process of predubbing.

As noted earlier, predubbing is simply making mix decisions in smaller, more manageable chunks. Instead of balancing an unwieldy number of tracks,

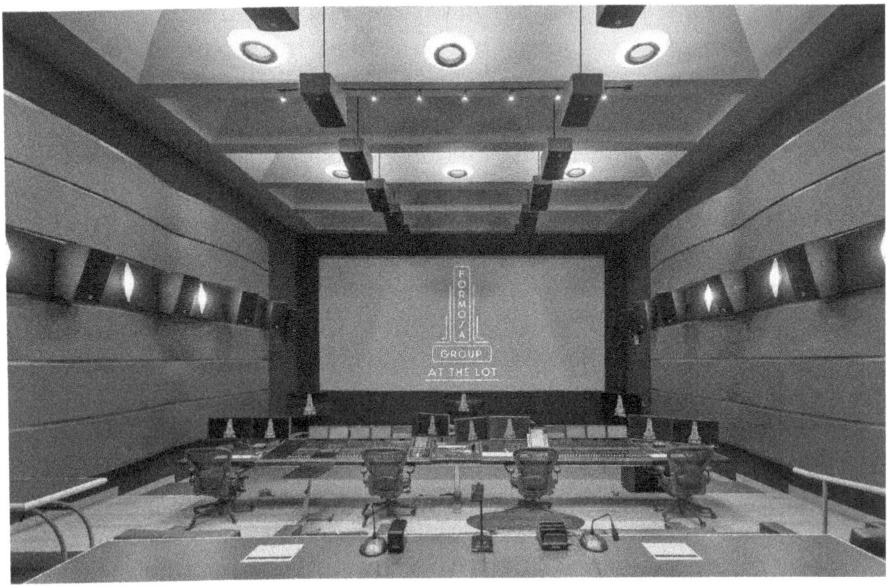

FIGURE 2.20 Formosa at the Lot Stage A

Source: Formosa Group.

the mixing team will tackle similar groups of sounds, lock them into place relative to each other, and then move on to the next grouping.

For instance, a scene might be comprised of Foley effects, conventional sound effects, ambiences, or backgrounds, and any number of commentative sounds. It could also have several music tracks going if the composer provided stems of their cues, and there could also be several dialogue tracks organized by character, interior or exterior, replaced dialogue (ADR), and the list goes on.

It's easy to see how daunting a task the mixdown can be.

In the analog days, there would be a limited number of machines to play all those tracks – physical tracks require physical machines to be heard, and all that machinery and media can get expensive. The predubbing process was essential to maintaining control over all those elements, and decisions would need to be made that could free up those machines for more elements down the line. Similar sounds would be combined, or mixed, together to create composite sounds, after which the next batch of sounds would be balanced, combined, and mixed down again. Dialogue and effects got the predubbing treatment, and as a matter of course, the music elements would typically arrive fully mixed.

The advantage was the optimized use of machinery and media, but the downside was that those elements had been committed to a printed track. If the balance was off or somehow not working against all the other elements, the final mix would have to be stopped while those predubbed tracks were rebalanced, reprinted, and then brought back over to the mix stage.

Judging how all the tracks would work together was difficult when they couldn't be heard until the final mix.

"Most mixers are working with the tracks virtually," says Mark Mangini, "meaning that we're not committing them to a recorded track." Digital audio workstations have enough processing power now that they can play hundreds of tracks at once, so tracks will be virtually grouped together and then represented by a single fader on the console worksurface. That one fader might represent a dozen or more actual audio tracks and effects, but having them balanced relative to each other within each grouping, the number of faders the mixer might need to access can be drastically reduced.

"You always have access to the source audio that created what you're hearing," continues Mangini. If a sound within a grouping needs to be adjusted, it's simple to open that virtual group, make an adjustment, and continue the path to the final mix. No pauses in the process needed.

The most significant benefit to the technology is that the balancing and pre-mixing previously relegated to the predubbing process can now happen at every stage, from editorial to the final mix. Mangini adds, "You couldn't create the kinds of soundtracks we create today without hundreds of tracks and the benefits of the technology."

The mix stage is an expensive space, and the ability to hear sound accurately in that space is paramount. It's not just about the technology and state-of-the-art speaker systems; the acoustics and sound treatments are equally crucial to the experience. The mix environment reveals all; any subpar or inadequate sound design will quickly reveal itself. This applies to all mix environments, from feature film mixing to studios producing streaming or gaming content and studios that do advertising work – in every case, the room needs to work accurately and portray sound honestly to provide the most accurate listening environment possible.

CHAPTER THREE

Creation – How We Do It

Applied Sound Design and Workflows

3.1. HOW SOUND WORKS – SOUND TYPES

The title of this chapter is probably a bit of a reach, as so much has been written over the years about the physics, the application, and the perceptions of sound – such a text could end up as epic as trying to write a comprehensive history of the world. The very idea that I could explain everything about how sound works in just one chapter – or even an entire, very thick textbook – is more than a little daunting.

Instead, the plan here is to provide some ideas and concepts as an overview, ultimately hoping to spur you on to greater heights of creativity in your approach to making sound work for you.

The central point – made many times previously in this text – is that organization is key to freeing the mind to go forth and create. Several points will be demonstrated here with some examples and reference materials; the hope is that they will facilitate your own application of the concepts. After all, no artist can create without ready access to the tools and materials they need, and the media arts are no different. For us sound people, our palette lives in production sound, music, the use of the Foley pit, and databases of sounds. Our tools are computer hardware, software, and sound equipment like mixers, speakers, and microphones.

Now that we have some foundation in sound, an overview of the process, and have gotten acquainted with the many players and their processes, it's necessary to corral those resources and plan for what happens next. The path forward lies in how to utilize all those moving parts best and harness them to work toward a final goal.

DOI: 10.4324/9781003373094-4

First, we must organize some thoughts and concepts. Some more sound theory, if you will.

As we move forward, it might help to think of all these terms and processes as the primary organizational tools for your track-building and sound editing practice to come, helping you further organize, define, and arrange your tracks by category, all to better set you up for the final mix. Knowing the sounds we have at our disposal is very important, and it helps to lay them out thoughtfully and logically, especially when working as part of a team. Someone else might need to edit or mix from those tracks, so organization and a logical workflow will help to facilitate that process.

It will also set us up for what comes after the mix as we prepare the final deliverables for distribution. That is an important step, and it further emphasizes the importance of implementing and following some best practices along the way. In the end, we need our collaboration to work efficiently and effectively. We want a stellar end result, both creatively and technically, and we need to keep our distributors happy by providing them with what they need to help our story find its audience and succeed.

3.1.1. Diegetic and Non-diegetic Sound

Starting out, it helps to identify and categorize sounds in relation to two basic sound categories: diegetic and non-diegetic. This is very broad, organizationally, but at least somewhat familiar in concept.

Diegetic sound is, very simply, sound that originates within the world of the visual story. Diegetic sound is attributed to the dialogue, actions, and setting we see on screen; these are the sounds the characters within any given scene can also hear, react to, interact with, and participate in. Dialogue, ambient sounds, and Foley would all be considered diegetic sound.

Non-diegetic sound is sound that comes from outside of that world. This is sound that furthers the storytelling narrative, yet it is sound that the characters within that story cannot directly hear or interact with. Non-diegetic sound typically comes down to two main sound elements, at least most of the time: narration (or voiceover) and musical score. As said before, so many rules in sound are made to be broken at times, so there will always be exceptions to this rule. Many films willfully break with that conceit – non-diegetic music can become diegetic (or vice versa), or there may even be some instances where the characters hear and interact with the narration, breaking the so-called "fourth wall." These are all creative decisions for storytellers, but they are certainly exceptions to the rule.

Most of this discussion will focus on the diegetic and how we can better understand the placement of the characters in their film world.

3.1.2. Categories of Sound

To help us further our seemingly never-ending quest for organization, we can additionally categorize sound by its mission in the film. While the following categories won't appear on any mixing cue sheet or sound editor's track build, they are helpful as yet another organizational tool when designing and building tracks.

3.1.2.1. Contextual Sound

These sound organizers apply primarily to the diegetic domain, where we first encounter contextual sound. The best example of this can be found in any situation with an onscreen sound source: contextual sound is the synchronous sound that goes with that action, most simply summed up in sound speak as "see a dog, hear a dog." Most production sound falls into this category as dialogue, which is, for the most part, synchronous. We certainly include synchronous sound effects, such as doors opening or closing, cars passing, and so on, with most of our Foley work also falling in this category. When that door slams shut on screen, it's up to our effects and Foley team to make sure that the proper door slamming sound is there, that it's believable and in sync with the action of that door closing. Synchronous sound effects are a potent tool in sound design, even considering the obviousness of their nature. Indeed, when we see that dog bark, we would also want to hear it. But what of the other incidental sounds that might go into a scene? We might not consciously miss footsteps, clothing rustles, and the like, but they bring an element of reality and believability to the scene that, when missing, makes the illusion we're striving after seem less real.

All sound plays a role in telling the story, so while a door slam might be what is observed visually, the many qualities and character of the sound we can apply help further its role and place in the scene. Is it placed tightly between, or perhaps even during, lines of dialogue? We might not have as much "play" in that case, where a bigger sound will have too much weight and possibly minimize or otherwise diminish or obscure the surrounding dialogue. In this case, the sound of the door would have to be of shorter duration and lower in level to fit the context of the moment.

Let's assume the door slamming is in the clear and free of any dialogue overlay. Is that door being slammed out of anger or frustration? If so, it helps to identify qualities we can infuse into that sound that will give it a quality or character so it sounds angry and gives the moment an added emotional weight. It helps to consider the door's construction – is it wood, metal, glass, or maybe some combination of those materials? What are the mechanics of that specific door? We should have at least an idea of the type of latching mechanism, but also consider whether its hinges should squeak, whether the door itself is moving across carpet or other flooring – there are so many ways to go, and we're just getting started.

(See *Media 3.1m – video, chess scene sound design treatment, Foley demo.*)

Foley is most often utilized as contextual sound; it helps to further the narrative by supporting the action on screen. With careful effects editing and effective Foley work, however, it can also work as a lens of sorts, helping to focus and direct the viewer's attention toward a visual cue or an action we might otherwise have missed.

In this media example, we have the premise for a potentially unexciting scene, where the only action is two people playing chess (see also Figure 3.1). However, we can heighten the tension of the moment by directing our viewer's attention to some very subtle visual cues. Our characters are both intensely involved in their game, but we realize that one character might have more going on internally than just being a part of this otherwise garden-variety chess match. We want to heighten the tension of the moment and emphasize that he is perhaps more deeply stressed by the way the game is going. By underlining his fidgeting with some well-placed Foley effects, we can help draw attention to that specific manifestation of his stress. That action was always there, but by choosing to highlight that action with sound, the viewer is also made more aware of his internalized stress.

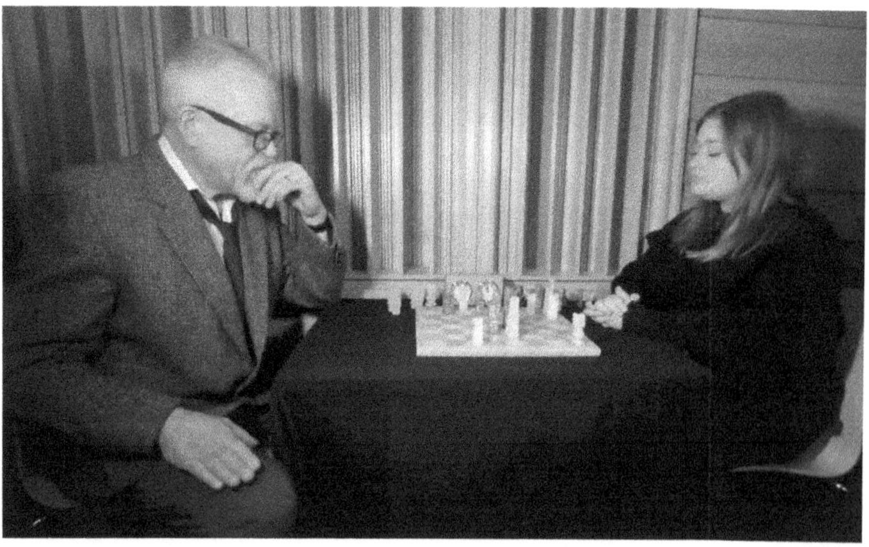

FIGURE 3.1 Through the use of Foley, we can draw the viewer's eye to the action we want them to notice. In this case, the fidgeting of the man on the left.

3.1.2.2. Descriptive Sound

Next is the category of descriptive sound. Sounds from this family can inspire greater creativity in our sound design approach, but there's more freedom because they're not directly synchronous with any action seen on screen. These tend to be ambient sounds, and we're developing sound-field types of sounds here, such as traffic, residential neighborhood sounds, birds tweeting, sounds in the background, and even room tones. While the direct sources of these sounds are not visible on screen, their addition helps to establish the setting through the implication of a specific geography, setting, and possibly timing – they can help add a sense of night or day, interior or exterior, and so forth.

At this point, it's important to emphasize that every sound added to the soundtrack is not just a placeholder or filler; it's an opportunity to further the storytelling, and the creative aspects begin to show themselves as we uncover the motivations of the scene. What is the mood of the scene? Happy or sad, somber or upbeat, winter or summer – the sounds we apply can help to impart those qualities, at least to an extent. There are many ways to go here, and we begin to inflict some emotional intent on our soundtrack by the sounds we choose.

(See *Media 3.2m A, Media 3.2m B – video, two sound design treatments, sound design demo.*)

In Figure 3.2, we see a riverside park in New England on a bright, sunny, early spring day. The setting is generic; it could be almost anywhere, with nothing exceptional going on visually. We have a sense of nature with some distant traffic, but nothing of note. It gives us nothing to cue any sort of direct sound, but there's still an opportunity to inflict some emotional content, so this scene has been treated in two very different ways.

The first treatment we'll call our "happy ambience," with the sounds of happy tweeting birds and some kids playing in the distance, with a reduced sense of traffic in the area. In the second, the sound design takes a darker turn, with more tension and discord in the ambience. Instead of happy birds chirping, there's the sound of crows and jays – a much more grating sound than the chirping heard previously. There's also a distant siren and a dog barking for an added edge, with louder traffic. The overall effect is that this area might come with some issues, and while this park looks OK, there's a sense that everything is not as good as what is being seen here. Perhaps it's a bit of an oasis, and we might be just one small step removed from some sort of turmoil.

Descriptive sound can be the literal sound that might be heard in that location or setting, but it also allows us to further the storytelling in the scene. All the sound sources are unseen, so it frees the approach to the sound design. If the story itself is furthered along by the sound treatment, with the sounds

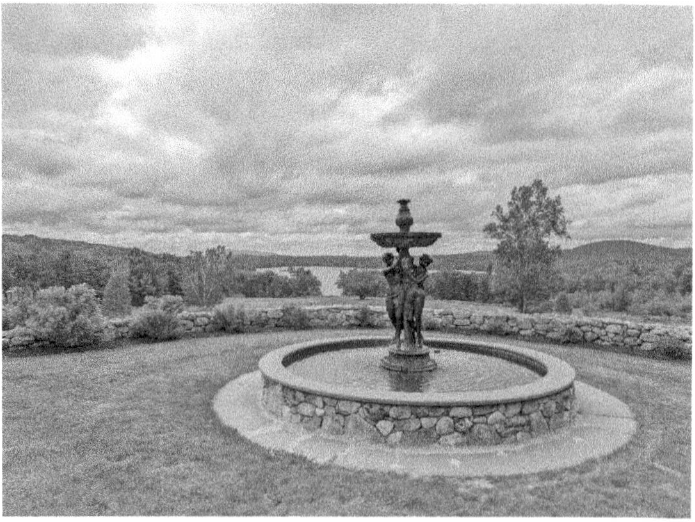

FIGURE 3.2 A waterfront park in New England. Different meanings come from different sound treatment approaches.

themselves explained or rationalized as plot points in some way, it all works toward the storytelling objective.

3.1.2.3. Commentative Sound

For our last category and organizational sound header, we have commentative sound. Sometimes, it's meant to convey an internal moment for a character we're seeing, but it can also be part of a broader treatment to help lend a scene a greater sense of emotion. The sound treatment can skirt the line between diegetic and non-diegetic, but it is typically devised as an emotional underscore to, or commentary on, the action we see happening on screen.

(See *Media 3.3m – video, a man walks onto a sports field, sound design demo.*)

The attached/linked scenario shows a middle-aged man walking onto an empty soccer pitch (see Figure 3.3). The sound we're hearing has almost nothing to do with what we're seeing visually; it's simply a sports field at sunset with no other people present. Instead, we see our main character looking around, perhaps wistfully, while we hear the distant sounds of a large crowd cheering. Much information can be inferred from this somewhat limited context, but the sounds we hear tell us that this moment belongs to him and that he's experiencing something emotional.

In this straightforward treatment, the viewer is given a particular insight into the emotions and state of this character. There's still much in question, and

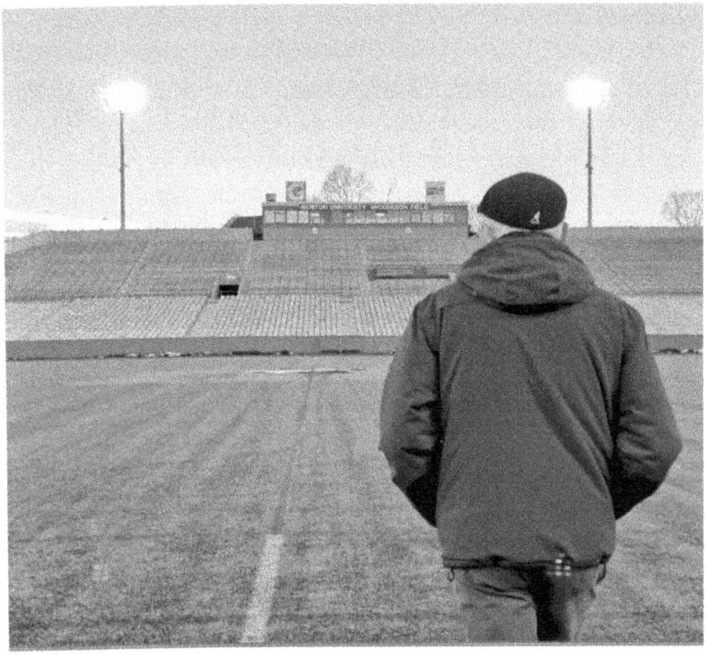

FIGURE 3.3 There isn't a lot happening visually, but we can use commentative sound to evoke the man's feelings of nostalgia. A distant crowd cheering can help bring out the wistful feelings our subject might be having.

without any context, there will be a lot of room for interpretation. The sound treatment is compelling, and it helps the viewer infer that he is having some kind of internal moment; perhaps he's reminiscing about his time on the field as a participant in an event that happened there. The sound design treatment is evocative, and it's likely to raise as many questions as it might answer, maybe more, bringing us wholly into emotional territory.

Working with commentative sound is a powerful tool; it can convey an emotional subtext but in a different manner than a music score. It also lends an emotional context in a way that might not be as easily accomplished with music. The sound is furthering the storytelling on its own, giving us some insight into the character we might not have gotten otherwise.

Commentative sound can also implicate an external influence, as opposed to portraying an internal moment. In those instances, sound design is being applied in a role more like a musical score, but the difference is that the sound somehow happens in the world of the film, influencing the characters and their plans. It could be a dramatically altered version of a natural sound or entirely unrealistic and abstract. This type of sound treatment is just another tool to help the narrative and interpret our characters and their development in the scene.

3.2. MAPPING A PLAN

One can begin to imagine that there are nearly as many ways to approach a soundtrack as there are practitioners of the sound design arts. There is no singular approach that will be the best (and only) path to getting the most effective soundtrack and sound design for a particular story. There will always be differing methodologies to get to that finished soundtrack, and it's up to the sound designer in us to determine the actual means to reach that end. We are all searching for that perfect recipe of sounds to help tell the story, and the type of story we're trying to convey will further inform our approach.

For example, one would approach an action film in an entirely different manner than a historical romance, and certainly, neither approach would work in a documentary setting. So much of this process is in the experimentation and figuring out what works best – and what doesn't. It's essentially trial and error, this process of finding and choosing just the right sounds for any given moment. The creative approach originates in the individuals doing the work, with the sound designer bringing their unique sense of aesthetics to a project, often accessing more profound parts of their psyche and sensibilities to help bring a soundtrack to life. We are working hard to put ourselves into that place and figure out how we ourselves might be motivated in a given situation – and then how to best articulate that idea or emotion through sound.

"I get hired because I have an opinion, and the director wants that opinion," says Mark Mangini. A kind of below-the-line typecasting might occur when a sound designer or sound supervisor becomes – rightly or wrongly – identified with a particular style of work or genre of film. That sound supervisor, in turn, will bring on people that they feel best suit the needs of the project as well. "I do a sort of a casting exercise," Mangini continues, "and I cast those hires based on their sensibilities. There are some editors or designers whose skill sets lend themselves more to horror versus romcom or to animation versus science fiction, and I will start making those calls based on what I think their skill sets are."

All this happens via technology, which facilitates all the creative personalities working in harmony with the gear, and in the process, it all serves to facilitate the creative vision. At other times, progress can be blocked or held up by technological shortcomings or a lack of know-how; sometimes, a sound treatment can be imagined, but there's difficulty bending the capabilities of the technology at hand to the creative need. At other times, there can be elements of both going on at the same time, with tension between the creative forces and the technical capabilities of the process. For the most part, however, technology does a fine job of keeping up with and accommodating our visions, and that means we can (and should!) keep dreaming big – pushing boundaries is how the art of the process is advanced and technological gains are made.

To get there, we need a plan to help with all these organizational aspects. While the creative parts and execution are, ultimately, entirely up to you, we will outline some pathways here to help get those creative juices flowing while also providing a path forward to a fuller, richer sound design experience. We will also use those organizational tools to parse better and organize the different buckets of sound we'll be working with.

3.2.1. Various Workflows

As sound supervisors, we can – and should – get into the process as early as possible, even being on hand during the development stage. "I've been advocating for years on the value of starting sound as early as the screenwriting stage," says Mangini, whose sound design work serves as a prime example of sound being a part of the story, not just an afterthought. "I've been engaged as early as the screenwriting stage on a number of films, and with great success."

Mangini continues, "There's one director who called me as he was writing the screenplay because he was trying to work through a narrative challenge and feeling stuck, simply because words are sometimes clumsy. He would ask me if we could solve this narrative problem with some creative use of sound versus the use of words or exposition. I don't always have an answer, but on certain rare occasions, I had an answer that helped propel a scene in a much more efficient way than the narrative could have done. Bottom line, I advocate for filmmakers being as open as possible when it makes sense to include sound as a storytelling tool."

It only makes sense to include sound as part of the storytelling process, so the earlier the sound designer can get involved, the better it will be for the story overall. Screenwriters aren't sound designers, and vice versa, but the collaborative effort that is visual storytelling is on full display here. Mangini's point is not unlike the old saying about how a picture is worth a thousand words; it's important to remember that an effective sound treatment might also speak volumes more than the spoken word ever could.

The sound supervisor could effectively work alongside the shoot, developing sounds and coordinating with the sound team on set with the production. When working on *Blade Runner 2049,* Mangini said, "Denis [Villeneuve, director] wanted to bring sound design on during production because, as a filmmaker, he understood the value that sound can have during the shoot, especially as they begin to edit.

"As we saw the dailies and early rough cuts, we would build sounds before score was even generated. We could cut sound that would indicate, or maybe even mandate a slightly different edit and then, once it got to returned to us, we'd improve the sound and the cycle would begin again. We created a very symbiotic circle of sharing." The partnership with Denis Villeneuve continued with their collaboration on *Dune* in 2021. "On *Dune*, we went even further to where the visual effects department had a presence on set, advising the shoot. So, as we developed sounds, those sounds would go to visual effects and actually influence the look of some effects before they were even animated. You can begin to see the value of sound being an organic player earlier than is traditionally recommended."

At the same time, the practicality of becoming involved at the production stage is invaluable to the post team. It's not meant to dictate the production sound department's practice in any way; it's more to provide background information for the sound supervisor. It can be hugely helpful to know, for example, that part of the shoot will happen on a live city street, a seashore, an airport, or any number of possible locations that would each come with a unique set of noise challenges. Sometimes, the sound supervisor can work with the production team to help mitigate those issues ahead of time, but more often, this meeting of the minds will serve a more informational purpose for the sound post team, simply by alerting them to the issues that will be waiting for them when the production wraps. Dialogue editors are better served when they have an idea of how much ADR might be needed. Sound effects editors can be tipped off to unique sounds they might need to come up with for their work in post, such as ambient sounds unique to a location, geography, or time of year; it also helps to anticipate any vehicles, machinery, or other set pieces that might be unique to the production. Sometimes, those set pieces sound nothing like we imagined they would, but starting from a place of accuracy first is essential.

Let's look at a specific example: a scene in a remote country setting in the northern United States in summer appears to be a remote wooded area, lightly settled, and on a large lake (Figure 3.4). There are several things to examine here, starting with the logistics of the location itself. First, did production happen in the remote geographical setting it appears to be, or was it staged in a stand-in location that was more convenient to production resources? Likely, it was within a commutable distance to an urban center and, ultimately, the filmmaking resources and infrastructure the production would inevitably need.

We see this often, where production will be staged in a location that isn't nearly as remote as it seems just because of the number of personnel, facilities, and resources needed to maintain a high-functioning working set. Our setting might only need to *look* like rural, remote Wyoming, but for any number of reasons, ranging from budgetary to logistical, production simply can't happen there. In these cases, the production sound department might be contending with issues like traffic and other less-than-rural noise issues, where what we actually need to hear in that setting is a pristine, remote environment.

After we've figured out a way around those ambient noise issues – primarily through careful microphone technique, possible dialogue replacement, and careful sound editing – there needs to be some consideration for the sounds

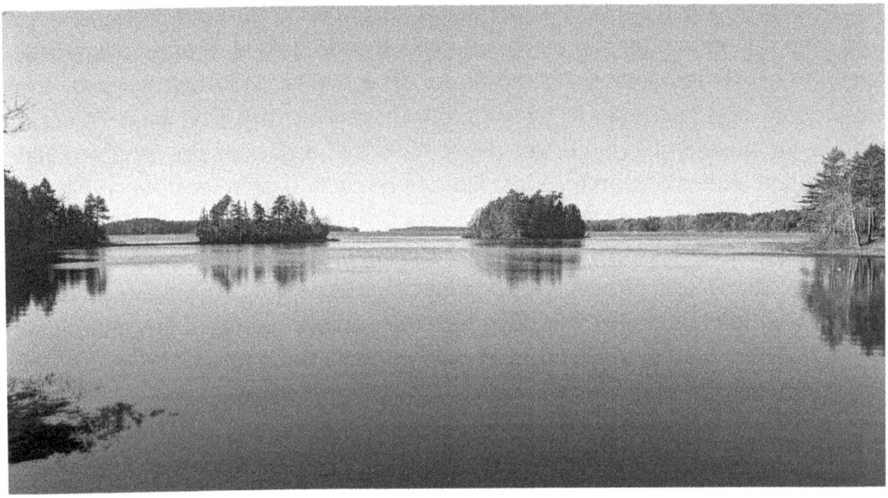

FIGURE 3.4 It's a clear day by the ocean, what sounds might we hear in this location?

that would realistically be going on in that setting and at that time of year, leading to the post sound crew creating a sound design hit list. There might be a lot of wildlife sounds that are unique to Wyoming in general, but there will be additional specificity imposed by the time of year. Wildlife, weather, and even those leafed-out trees probably would have a different sound character than they would in the fall or winter, and a lot more avian life will be present, from water birds to birds of prey, making it important to ascertain what species dwell there in the summer. Once there's a baseline of descriptive sound, the sound team can start constructing the soundscape for that scene.

The location mixer can also help the post-production dialogue team if some noise issues can't be overcome on set. Let's assume the location is more accessible than it looks and is near a major highway. In a complicated, noisy location like that, there's sometimes little to be done about those noise issues. The location mixer might be able to arrange some time with the actors at a quieter location nearby and have them redo their lines wild – meaning without cameras rolling. This allows the actors to hone their performances while they're still close to their characters and in the right mindset, and able to play off their fellow actors – and ultimately get cleaner dialogue than they could have out there on the noisier set. We keep saying the mixer's task is to get clean dialogue, so the workaround, in this case, is that we appropriate our actors and create a less-noisy version of the track. In this case, we can do it without the picture rolling.

Bottom line, it's helpful to have a line of communication with the production sound crew – again, not to try to manage them, but to manage our expectations before the material comes to the mix.

Post-production sound personnel job descriptions can get highly specialized, and editors are usually assigned tasks by category and specialty, particularly on bigger features. It's common for an editor to be known best by their latest work and assumed to have greater expertise with that type of sound editing. As an example, an editor might have had a successful tenure editing artillery and explosions on their last film. As a result, they might become typecast as the person to call for those types of films.

The work will typically be organized into the sound macro-categories of Dialogue, Music, and Effects. These lanes help define the divisions of labor, but they're also the categories that align with the content needs of the distribution master. When the final soundtrack is delivered as *stems* formatted in this way, the distributor can provide the elements needed to recreate the mix from its core elements for international distribution.

This serves several purposes, the first being the need to re-dub the film into local language. On the distribution end, it's simple to align the three stems – Dialogue, Music, and Effects – and simply mute the dialogue track so local, native-speaking actors can go into the dubbing booth and replace all dialogue in the film. The second reason for this deliverable requirement is archival – having these elements separate from each other and synced with the master, we have a fallback option if the film is ever revised or re-edited. A film or program will often be re-edited and reformatted for different broadcast or streaming outlets, airline play, and other distribution channels.

In those cases, music rights also become an issue – each piece of music cleared for domestic theater release might not be cleared for international distribution, depending on the artist and the publisher. Having the flexibility of a separate music stem gives the distributor options with uncleared (and sometimes unclearable) music cues, making them easier to replace after the fact.

In this way, the sound build is approached in a multitiered, multifaceted way, and the sound supervisor will "crew up" or hire sound editors for the team, bringing them onto the team by specialty. The different editing groups will work separately but usually concurrently, arriving at the mix stage with their track groupings that, ideally, mesh together and work toward completing the mix. Communication is critical for all the teams to get there together with tracks that work together perfectly.

3.2.1.1. Dialogue

Dialogue editors are a breed apart; they excel at their specialty because they can find joy in it, and consequently, they're very good at it. It can be tedious work because of how exacting it is, and only the most detail-oriented need apply.

The challenge in effective dialogue editing is that every hearing human knows what good, articulate dialogue should sound like, so the moviegoing, story-watching public will be the final arbiter of how good their work is. This is simply because every hearing human is intimately familiar with the human voice; it's our primary means of communication, and being able to spot flaws and irregularities in any vocal material is deeply ingrained in all of us. Suppose the voice being represented is distorted, inadequately recorded, or "off" in any way. People will notice – sometimes only on a subconscious level, but it could be enough to break the illusion that good visual storytelling is attempting to impart.

The sound editor's maxim is "good sound editing happens when no one notices," and the sentiment is in full effect with any dialogue treatment. While the moviegoing public is unlikely to compliment the dialogue editor's work, they will immediately notice if anything is amiss and the dialogue track doesn't sound natural. This challenge is at least part of the appeal to some editors, and those specializing in dialogue editing thrive and revel in that challenge for any number of reasons. It can be a gratifying task when done well, and many dialogue editors appreciate it for its precise and exacting nature.

In terms of the overall sound workflow, it is necessary to have a preliminary dialogue edit as soon as the picture edit is locked so the needs of the dialogue track can be better assessed (see Figure 3.5). From this point, the sound supervisor and supervising dialogue editor decide which lines need to be replaced, either with alternate takes or through the ADR process. ADR is short for Automated Dialogue Replacement, a process sometimes referred to as looping. Much of the urgency of the task is in part about scheduling our actors for their ADR time if needed, as they're likely busy on other projects, making their availability a significant factor.

Dialogue is vital to the narrative and essential in the exposition of the film. Any sound professional will likely have some powerful feelings and opinions about how that dialogue element should be presented in any visual story. Some feel that every syllable should be articulate and just as clear as possible – after all, the last thing a dialogue editor wants to hear from an audience is, "What did they say?" While it's good to question the norm, push boundaries, and occasionally throw convention aside, the spoken word is most often our best guide to understanding the story. Parts of it going missing or getting obscured by other sound work will cause some controversy among viewers.

Several factors influence dialogue intelligibility in modern storytelling. Some of it could be performance-based, where the actor made the conscious decision to speak softly for whatever reason. Sometimes, it can draw the viewer in, while at other times, it's just annoying as the voice lacks clarity.

It can also be a creative choice that the director and the post-production sound team made collectively. Perhaps the balance of the film does *not* hang in the balance of every syllable of dialogue, and it's OK if some of that voice work becomes obscured by other elements – Christopher Nolan has done a lot of work in this area, but not without controversy.

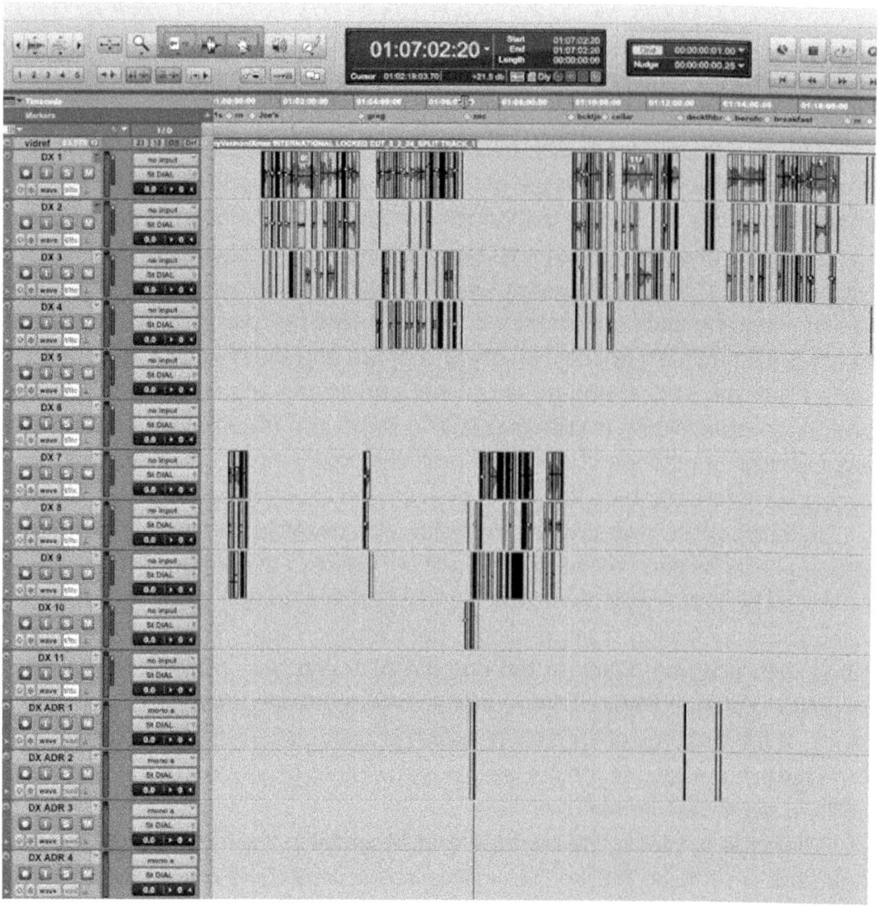

FIGURE 3.5 Dialogue edit in Pro Tools. The groupings of tracks are by interior, exterior, with ADR and futzed tracks at the bottom.

Source: Permission AVID Corporation.

We humans are tuned into the sound of other humans' voices, and when they are hard to hear and obscured by other sounds, it can be uncomfortable for the viewer, even to the point of being off-putting. Our ear wants to follow that dialogue and will instinctively work very hard to understand the voice element as it becomes more and more obscured. When that thread of narrative conversation falls away, it becomes problematic for the viewer, who can feel adrift when the dialogue becomes too difficult to follow.

Ultimately, giving the dialogue element its due is essential, and being aware of the weight it carries in the narrative is necessary for the exposition of that narrative. While the best sound work should not (and doesn't) call attention to itself, the challenge here lies in the fact that the average viewer can be incredibly discerning when it comes to the voice element of the visual story.

3.2.1.2. Sound Effects

Sound effects are organized by type, category, and source – and, at the same time, by any and all other means that best translate to the needs of the soundtrack at hand and the sound supervisor's preference. The layout can become very granular, but starting with some basics is best.

First, ambient sounds often go hand in hand with the dialogue edit, as matching ambiences and room tones to what was on set on any given day is paramount to a seamless dialogue edit. Our ambient sound effects will help build the scene's environment while masking any differences between lines of dialogue. Remember that dialogue might be sourced from many different takes, likely shot at different times and from different angles. The dialogue edit might also include wild lines and – most challenging– newly recorded replacement lines, or ADR. The design of the ambient effects is the sonic foundation of the scene, a constant that helps the dialogue sound and feel consistent throughout.

The sound effects edit will be further organized by the categories of Foley, hard effects, and soft effects, each lending another layer of organization for the mix later. For example, when mixing, it's helpful to know where the hard effects, such as a door slam, might appear in the track build, as opposed to sounds with a softer attack, like a car driving by or the surge of a crowd applauding. Mixers don't like surprises when dealing with unfamiliar tracks, so a logical layout and exposition are essential practices. While different editors will have different styles, methods, and modes of editing, mixers also have differing needs on their end of the process. In all of this, it is up to the sound supervisor to lead the charge, establish an organizational template, and impose a type of style guide on the track build that will make the most sense for all members of the post-production sound team down the line.

Sounds can be further organized and divided by type or family. For example, mechanical sounds, such as vehicles and all their moving parts within; sound people call that a workup, where a recording team has acquired the vehicle in question and recorded all the sounds that the vehicle can make in all modes of operation. These would include engine sounds, idling and revving, pulling away, arriving, driving by at varying speeds, and so forth, but also all the mechanical sounds the vehicle might make, such as doors and hood opening and closing, suspension squeaks and rattles, switches, buttons, and pedals – the list can go on.

Another complicated workup would follow anything to do with horses, where the editor will need to separate hoof sounds and all the ways a horse might move, from a canter to a gallop to a trot, in addition to all the different vocalizations a horse might have – whinnies, neighs, and snorts – and the tack they are wearing, which would include saddles, leather harness creaks, buckles and stirrups, and so forth.

Sound editors have an uncanny ability to hear *inside* a sound, analyzing it and then dissecting it into elements that can be more easily reproduced, combined, and manipulated to create a rich and realistic version of that sound. That dissection can result in a hyperdetailed sound package that can take up a lot of track space, and this is just a part of what will be produced.

Commentative sounds – those internal moments discussed earlier in the chapter – also get their own track set so they can be treated differently when they occur.

It is important that, in all cases, there be some discussion between all the sound editing parties, particularly the sound effects crew, to lay out some guidelines, establish a style guide or standard operating method, and find an approach that will work for the whole team. Figure 3.6 is one example of how tracks might be organized for the members of the post team. This ensures that everyone has a directive, they all get what they need to succeed, and that all those expectations are effectively and efficiently managed. Adding a high level of coordination to the sound crew's operation is yet another of the sound supervisor's list of tasks; they're the sole individual tasked with communicating between all parts of the sound crew.

All sound work in post-production should strive for reality, or at least a kind of reality that best suits the world of the visual story. One of the most effective tools in achieving realistic sound in this contrived world is that of the Foley artist. A good Foley person will add a dose of credibility to a track that pre-recorded sounds would not be able to attain. As an example, we might be able to use some perfectly serviceable footstep sounds taken from a sound effects library or even footsteps that were recorded wild at a different time, but the reality is that, for many reasons, those sounds often won't match the nuances of the action being seen on screen.

As a simple example, consider how a person might walk into a room. This is a common action that happens every day but is never as simple as just "left foot, right foot, left foot, right foot…" up until the moment a person stops walking. This is often how walking sound effects are presented in any given sound effects library, simply because it would be impractical to include every possible scuff and variation possible. In a real-life example, there might be some off-rhythm footstepping as that person moves through a doorway, the surface being walked on might also change as they transition into the new space, and a person almost certainly doesn't arrive at a stopping point by simply not walking anymore.

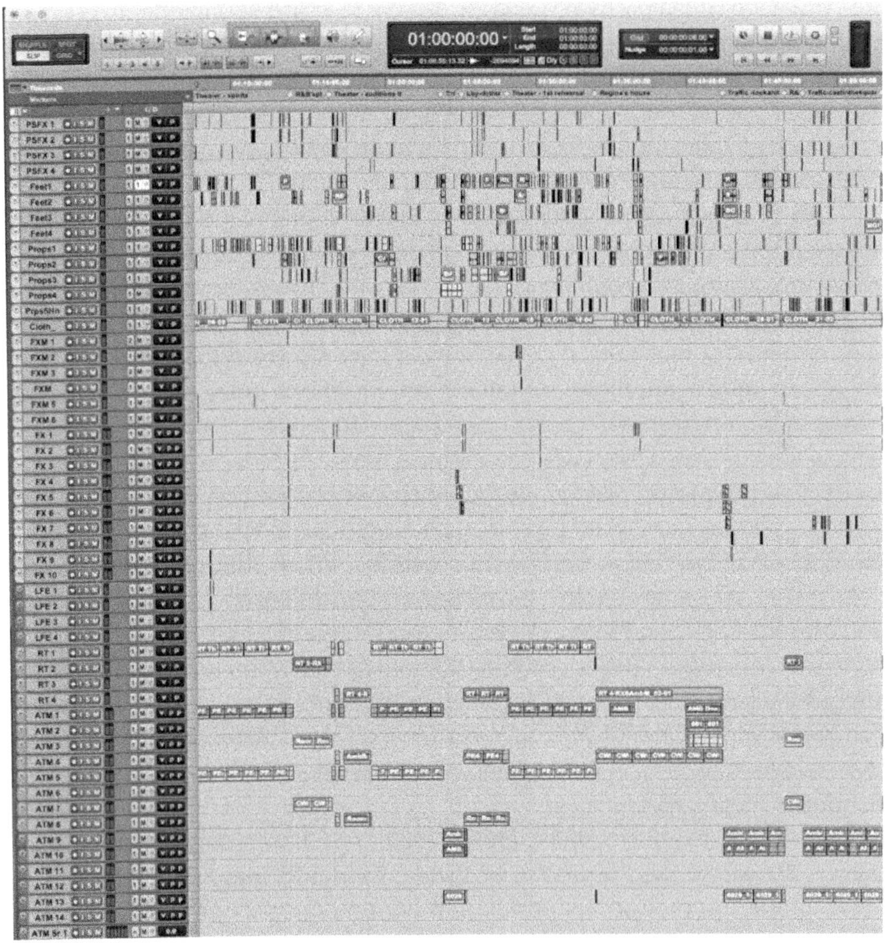

FIGURE 3.6 SFX tracks in Pro Tools. Note how the tracks are grouped by type – Foley, hard effects, ambience, and so forth.

Source: Permission AVID Corporation.

Usually, a person will slow down, maybe scuffing their feet as they come together and turn to face other people in the room, shifting their weight from one foot to the other as they do so. While subtle, the realities and nuances of these actions can be challenging to replicate from any sound effects database.

Taken to another level, one begins to realize that it's not just about feet – Foley also accurately presents and conveys the sounds of props handling and any other physical actions, movements, and clothing rustles – essentially any action being taken by the actors on screen. Foley is a far more efficient means to creating these sounds, and in a far more convincing and compelling way than hours of sound editorial alone could ever hope to achieve.

There are ultimately two objectives at work in all the sound effects building. First, there is a need for a soundtrack that supports and furthers the story, where every sound applied works toward furthering the story. Every sound effect has a part in moving the story along, so every sound represents another opportunity to help tell that story. It can be a painstaking process, selecting the right sounds to represent what's being seen on screen.

The other goal we must remember is that all these sound effects elements still need to be present and have enough detail when the dialogue is removed. When the distributor is preparing the film for other markets, that film will need to be redubbed into that market's language and must be present with a high level of detail for that version of the story. It may require additional work from the sound team to get there, but the bottom line is that every action on screen needs to be represented in sound for foreign distribution. This detailed sound effects track is called the "Fully Filled Effects track," or FFE.

In that production dialogue track, when it is edited, non-dialogue sounds from the set may or may not be able to be saved, but even if they are, chances are good that even more detail will be needed. When that production audio track is removed, it takes with it the natural and incidental sounds that help hold and maintain the scene's sound integrity and believability, as well as the atmosphere of the scene. This is precisely what happens when providing separate stems for distribution, particularly foreign distribution. As the story is distributed in markets with different language requirements, each distribution channel will likely require that story in that market's local language, so it will need to be redubbed by native speakers.

In general, the entire sound effects team, but more specifically the Foley team, will need to step up and work toward executing that Fully Filled Effects track, or FFE, to provide the illusion that we're hearing those actors speak a different language, and the proper execution of all the incidental sounds will help sell that concept.

There has been much debate on the subject of dubbed vs. captioned films presented here in the USA and which version of a film might be more marketable, but the tastes of the American filmgoing public won't be debated here. The fact is that many international markets have their tastes and precedents for this, with some regions preferring dubbed local language and not caring so much about lip sync. In contrast, others prefer to read subtitles and hear the original actors' actual voices, performances, and emotions as they were originally intended. There are merits to either approach, but the bottom line is it can't always be predicted how the story might be received and consumed. It's always the best practice to provide for all scenarios from the beginning by helping to set up the distribution channels as well as possible and providing those mix stems.

Very often, the process of building and mixing the FFE tracks can lead to a sense of déjà vu, and it can feel as if the same film is getting mixed twice – initially for local distribution, as intended, and then again, without the dialogue track and including the FFE for international distribution. Like many solutions, this one also comes with its consequences, the main one being the loss of some incidental sounds and ambiences when the original production and dialogue tracks were removed. Here, the sound effects team will strive for a seamless, credible, and supportive soundtrack that fully supports all the on-screen action while being believable.

Another consequence can become apparent later in the process when the timings of the translation differ from the original, sometimes by a significant degree. Making those timings work for those markets is a major challenge for the translators, who will need to adjust phrasing so that the performers don't need to rush. In all cases, a certain amount of remixing will be necessary, so music and effects tracks need to be presented in a less dynamic way than they might have been in the original.

The track required for foreign language distribution will appear in the deliverable schedule as an "undipped" Music and Effects (M&E) track, simply meaning that the ups and downs of that track in the mix that so neatly fit around the original language need to be undone, at least to an extent. For these deliverables, a mixer will have to spend some time remixing the soundtrack – essentially flattening out the peaks and valleys in the soundtrack so that in the internationalized version of the mix, the sound effects and music stems can be adjusted to better conform to the timings of the translated dialogue track instead.

3.2.1.3. Music

The concept of sound design is intimidating to many; it can be a mystery, and otherwise rational people often don't even know where to begin. However, working with music frequently befuddles them to an even greater degree. Perhaps the ethereal nature of sound makes it hard to wrap one's head around – and music seems even more ethereal, with the concepts of music theory and instrumentation and arrangement coming into play. Surely, one must be some sort of master with mystical powers to even contemplate working in this field!

Fortunately, it's not that hard, and just as with all the other areas of sound and sound design, it's not hard to find a professional to interpret your tastes and instincts and channel them into coherent music choices.

Music can occur in many modes in visual stories and come from many disparate sources. In an ideal scenario, a composer will collaborate with the director to create an original musical soundtrack to accompany the film. The composer can then tailor the score to best fit the action and story on screen, orchestrating pauses for dialogue to poke through, emphasizing and accentuating actions, and providing the emotional undertone to the work.

The music score can also be sourced from prerecorded compositions, such as classical pieces, pop songs, and other produced music, chosen for their emotional and visceral impact on the visual action. Music can also be sourced from music libraries, which are music-for-hire databases often administered by large music publishers. Here, a storyteller can license a music cue for a fee or even a whole suite of music cues for use in their visual story. Where and how the piece will be viewed typically dictates the rate structure for licensing. As one would guess, a widely distributed film will likely elicit higher licensing fees than a piece with a limited, online-only, or festival release schedule.

The music department might require several skill sets, depending on the scope and scale of the work. In cases where an original score is desired, the composer is at the top of the list, and a music supervisor may be brought on when parts of the score are being sourced from pre-recorded works, depending on the needs of the film. The composer might bring in a team of creative and support people, including arrangers, orchestrators, producers, engineers, editors, and, of course, musicians.

A music supervisor often presents as a walking encyclopedia of musical knowledge; they're incredibly knowledgeable about music styles and artists and know where to find a piece of music that perfectly fits the moment or theme. Their role is that of curator, so it is less personnel-intensive than the music department behind a composer, but they do require some administrative help. Their team consists of people versed in the legalities of procuring or "clearing" the rights to any music they might choose; they ensure that all rights and clearances have been procured and licensed, meaning legal permissions have been granted from the owners of those rights so the songs can be used in the story for its final release.

The creative process varies with the composer. "In my ideal world, I want to talk general concepts with the director right away," says composer Ed Grenga (*Ghost Light, The Legend of Lucy Keyes*). "I want to get inside the director's head and know what they're thinking. Then I want to read the script."

As different composers have their own approach, the main objective is to connect with both the director and the script and help bring that vision to life. Mr. Grenga feels that early involvement is better for everyone: "After I've read the script, I'll talk to the director again and ask more questions about it. At that point, I'm already starting to write some music based on what I've read; I just want to jump on it and get started."

Just as a script goes through several drafts and iterations, the score will begin to evolve, taking shape to fit the film. In Ed Grenga's case, he'll sometimes work concurrently with both production and post-production. "I want to be involved at the very earliest opportunity," he says, "Basically, I do a lot of pre-production work, just exploring my own imagination. Then I start watching the rough cut, and I already have my own library that I created for this film." As the story moves through the post-production process, he will begin to piece some of those elements together, refining the score as it takes shape and comes into its final form.

All the work being done on the music side, whether composed or curated, requires the skills and services of music editors. A good music editor brings a solid musical skill set and sensibility to the work – they can read a musical score while taking advantage of the latest technology to edit the music cues into place, ultimately making the music fit the action we see on screen. Even an original music score can benefit from the services of a music editor – there may have been changes to the picture cut that alter the motivations in the music track. Changes sometimes occur after the composer has moved on to another project, requiring the music editor to work with the composer's original tracks to tailor them better to the new cut.

Ultimately, all music will be delivered as separate elements. These track groups, or stems, allow the mixer to independently adjust many elements within the cue and provide options and choices for the final mix stage. Just as dialogue will need to be replaced for international markets, some music cues might also need to be removed, having only been cleared for one market and not others. Broadcast, streaming, and theatrical releases each have different rules and licensing fee structures, meaning negotiations for the music clearances required for domestic and international use are specific to each release and piece of music. A cue may need to be replaced while not affecting the rest of the mix or the integrity of the scene or scenes it plays in. This makes another case for providing those separate DME stems for distribution.

By breaking out our deliverables this way, the marketing department can also pull shots and clips from the film to create trailers, promos, and commercials. By providing a dialogue track that is clear of any sound effects and music tracks, the marketing department can better build an effective promotional piece.

3.2.2. Following Through – Transitioning from Production to Post

Once the production crew has wrapped – sometimes even before they're done – post-production will ramp up, and the picture editorial department can begin assembling the picture cut. As always, it's helpful for the Sound Department to have an early line of communication with the editorial team, getting across the information needed to help them know what's happening at this stage. Early in the process, that knowledge is used to help with preplanning and expectation management purposes, but Sound can also be there to lend a hand if needed. A good relationship with the editor can resolve issues before they become real problems. For example, an editor might check with Sound to see if an imperfect or noisy dialogue track can be effectively repaired. Another example is when editorial needs specific sound effects, sometimes needed as placeholders to help with the story's timing. To plan for the pause required for a thunderclap, a car passing, or any event where the action on screen needs to pause for a sound element, placeholder sound effects help visualize how a scene might play out. At this stage, effects placement is typically intended to be a guide, understanding that those sound elements will be redesigned and replaced later by the Sound Department. In either case, those effects help tell the story and measure timing, pacing, and mood; they are essential to realizing the final version of the edited story.

It should be pointed out at this point that picture editors are, in fact, de facto front-line sound designers as well. Our new world of media and streamlined workflows means the lines between many job definitions can get blurry, and creating a coherent visual story involves considering sound at every level. Some of the best picture editors are, by default, highly accomplished dialogue editors and sound designers on their own. After all, the spoken word is an integral part of storytelling, and using sound well at all stages is essential to telling a compelling visual story.

Concurrently, the Music Department may also get involved at this early stage, doing similar work to help the story develop by providing placeholder music cuts or even outlines of the actual cues for the editor to work with and incorporate into the edit. As with the sound design, it will ultimately all be replaced by more polished and final versions for the final mix, honed and refined as the work continues, but this preliminary effort is made to help develop the story and move the progress of the edit along.

Most of the audio post crew won't begin their hard work until the picture cut has been "locked," meaning all changes and notes have been addressed, and everyone with a creative stake in the process has signed off on the cut. At this stage, it's time to stop messing around with the edit; it's a real show now. The reasoning here is severalfold, but mainly, it has to do with the other downstream processes like color grading, computer graphics, and, for the Sound Department, the real work of sound design.

Picture changes are relatively easy to execute for the editorial team, as there's essentially only one timeline, or "track," meaning only that when you add or remove shots, the rest of the timeline ripples to accommodate that. In sound, we can have several dozen tracks going across several departments. Dialogue, effects, ambiences, and music are all taking up a lot of real estate, tracks-wise, and not always happening in the same place at the same time. The result is that the simple ripple in the picture timeline isn't so simple to accomplish on the sound side of the workflow – the most straightforward of edits would need to happen across several tracks' worth of timelines and several departments. CGI and color correction will also have to contend with picture timing changes, with every new edit potentially introducing new errors into the workflow. For the sake of simplicity, it's better to lock the show than to have the various downstream players absorb the time-consuming processes involved when changing the timing of so much hard work after the fact. The bottom line is any editorial work after the fact needlessly consumes additional time in all the downstream departments – time that could be constructively put to better use. Time is money in the world of visual storytelling, just as in any other, and wasting money is hard to justify when simply planning better could have headed it off in the first place.

Once locked, the separate teams embark on their tasks and begin their work in earnest. There's a lot to do here, and the time frame and budget can greatly influence how it will all go from here.

3.2.3. Working in Sound Editorial

The first priority for sound editorial is getting those dialogue tracks in shape. A dialogue editor may seem redundant after the picture editor has already placed that dialogue in sync with the picture. At this point in the process, the picture edit dictates how the story will be told. So, the question is – if the picture is locked and the dialogue is in place and in sync, what is it that a dialogue editor even does?

Once picture has been committed to a locked cut, the first step is to clean up and organize that production dialogue track. This is where the dialogue editor will remove every sound that isn't dialogue while parsing dialogue to individual tracks, assessing the quality and viability of that production track. The time spent in this stage can vary, depending on the dialogue editor's and the mixer's requirements.

Typically, the tracks are organized by actor and interior or exterior locations. If there's a need to process some voices differently – such as through a telephone or radio speaker – an additional set of tracks will be set up for specialized signal processing. These are typically called "futzed" tracks, where various processors are called upon to make the track sound like it's coming through a phone or a speaker.

If there's any chance the effects team could use some of the non-dialogue production sound, those sounds are shifted over to a set of tracks to be shared with them. The dialogue tracks are tightly edited to consist of just the spoken words, exhales, coughs, laughs, and other vocally oriented sounds that can be used from production. There may be some gray area here, and different productions' best practices may vary – so in some cases, coughs, breaths, and other sounds that aren't vocalizations identifiable with any one character will all be assigned to the effects editing team, who will add them to the sound effects tracks. Non-vocal human sounds, such as breaths, kisses, and whistling sounds, are typically left to the effects team. While they may originate from the mouth, they're not language-based and don't engage the vocal cords – these sounds will be part of the sound effects stems and available for inclusion in the soundtrack for all the international distribution channels.

Part of the dialogue editing process will require the editors to fill in the room tone and set atmos, where necessary. The ideal they strive for is to make the dialogue edits as smooth as possible by making the background as consistent as possible so that there is no detectable difference or drop in presence in the transitions from dialogue to the spaces in between. Straight, hard cuts won't work here, and the editor will need to fade in and out of the lines of dialogue to make the transitions as smooth as possible. Otherwise, the viewer would hear the cutting points. Ramping in and out of those dialogue lines makes the cut point less apparent – and the first rule of sound design, after all, is to not call attention to what's going on with the sound.

As editors continue to clean up the production dialogue, they will also assess the quality of the track. Using noise reduction tools, such as plug-ins made by Waves and Izotope, gives the editors a lot of latitude in how those production audio tracks will work. It can make good-sounding tracks sound even better, and even rescue tracks that would have been tagged for replacement in the past. However, there will be takes that prove to be beyond repair, as well as script adjustments that need to happen after the fact, leaving the dialogue team with only a couple of paths forward.

Option one means going back through the takes from the production and seeing if any other takes are in better shape, sound-wise. However, they will also need to work with takes used in the locked picture, and getting the sync right between performances can sometimes be a tall order, especially if using different audio on a tighter shot. If the location mixer knew they were in a challenging situation, noise-wise, we can hope they were able to get some wild takes from the actors – only off-set and away from the noise and recorded by the actors while they were still inhabiting the role and the right frame of mind. But again, luck has a lot to do with it, as does timing, and sometimes it's just not possible or practical for the mixer to capture additional takes after the fact.

Option two, then, is ADR, or Automated Dialogue Replacement, a term coined by the Magna-Tech company and their line of mag film dubbers – the company is long gone, as is the practice of using linear physical media like mag film, but ADR as a process has been around long enough to see the term genericized. The process is also known as "looping," which came from the practice of cutting a physical loop of workprint film for the scene or line in question. The two terms are used interchangeably but mean the same thing in the context of dialogue replacement.

In addition to technical and noise issues, there are other reasons for resorting to replacing dialogue with the actor after the cut has taken shape. The performance itself might have missed the mark and isn't up to the context or the moment, and the director feels that the scene could be strengthened with a better delivery. This is reasonable, as actors often shoot scenes out of sequence, and maintaining continuity is always tricky in the best circumstances. Finding the right performance becomes even more elusive when a scene is broken into segments and shot over several hours, days, or even weeks.

New dialogue is sometimes needed to help clarify a plot point or storyline, or simply as emotive "sweeteners" or the reactive aspects of a performance, like breaths and sighs or sounds made in reaction to an on-screen action, such as in a fight scene. These are less taxing for the talent, as sync is usually more malleable than it would be in a tight shot of a conversation, but such nonverbal sounds can go a long way toward cementing a more believable performance.

The concept of dialogue replacement can be somewhat polarizing, exacerbated by the fact that it's also a junction where all the tensions between creativity, technical concerns, and budgetary matters collide. On the one hand, there are many directors, producers, and dialogue editors in favor of the process, while on the other, there are at least as many actors who dread and despise their ADR sessions. The actors have a point here – the sessions can be arduous and challenging on both a technical and creative performance level, and the act of performing in the sterile environment of the studio while rising to the emotion and level of delivery required by any on-set interaction is always going to be difficult, at best. Factoring in the technical issues associated with synchronizing, editing, matching the microphone model and type, and forcing all of that to work seamlessly with the performance being seen on screen, the result of this equation can mean some long days on the ADR stage as part of the struggle to find a match with the original performance.

A famous actor once argued against the process with the rationale that the right performance was the one that happened on set with the other actors while interacting with them and feeding off their emotion and energy in the same space and time. With so many variables and conditions, a performance's less tangible qualities can be challenging to capture again later. These are far less-than-ideal conditions for acting, with the actor isolated in a sound booth and under a virtual microscope, most often weeks (if not months) removed from the time when, and miles from where, the performance originally happened. The actor has very likely moved on to their next role and is in a whole different mindset, making it even harder for them to mentally get back to that place.

Some directors were – and still are, surprisingly – famously opposed to the idea of replacing dialogue at all. Robert Altman comes to mind – he would famously exclude any references to dialogue replacement in his actors' contracts; he felt it was something that would only detract from the performance he captured on set. He had a point, as he preferred freewheeling, improvisational dialogue over the rigidity of sticking to the script, and he felt that the technicalities of being under any post-production microscope would only serve to stifle that flow. He reversed himself later but remains one of the more prominent voices to have taken such a hard stance on principle.

In the end, dialogue replacement should be considered as just a tool, and only one of many that sound designers can wield in the search for a perfect soundtrack, but it should also be viewed as a process that can help support the creativity of the storytelling. Ultimately, it's not simply a blunt tool but a precise and creative instrument that can help preserve and serve the story better.

All paths lead back to the location mixer again here, and it should make apparent that it's not just helpful but imperative to get clean dialogue at the outset, as anything less is a compromise.

A typical ADR session will have the cues lined up and loaded into the workstation in advance. Different methodologies are available in workstations, systems, and plug-ins that take different approaches to this, while actors and ADR mixers have their individual preferences as well. At the core of the process, the actor needs to see the scene with the line(s) intended for replacement, with a system to cue them when to start with their new line (Figure 3.7). They can then re-perform the line in sync and with all the nuances of the original performance while the sound team captures this new performance in a way that matches

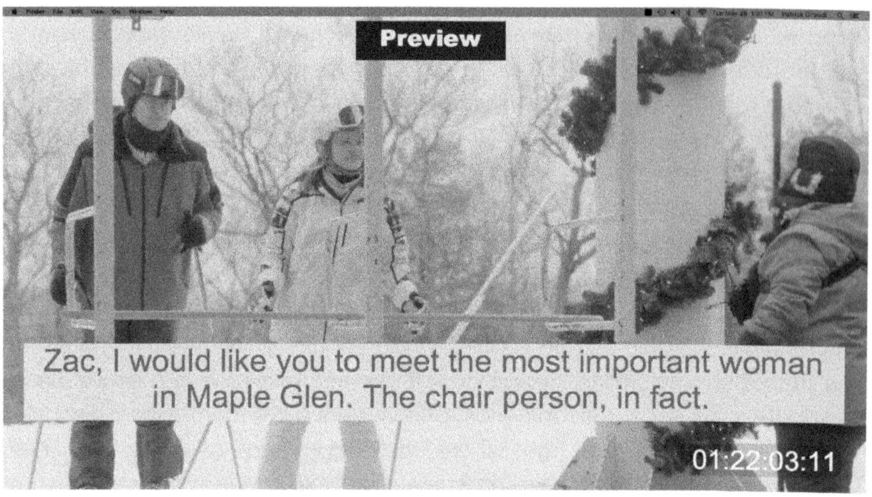

FIGURE 3.7 ADR booth setup – what the actor sees when replacing dialogue. The lines appear on screen while an audio and/or visual cue system tells them when to begin.

Source: Permission H9 Films and Moody Street Pictures. Photo courtesy Mark McBryde.

the original as closely as possible. Hearing the original line from the production track is helpful as a reference, but actors sometimes will want to have that muted after a couple of run-throughs or takes. The cueing system can be a series of audible countdown beeps, with a 3, 2, 1, and a silent 0, placed where the line would start – or it can be a series of "streamers" or flashing lines inserted into the video, culminating in a final flash at the center of the screen signaling the actor to begin. Some systems utilize both modes at the same time – beeps that synchronize with the on-screen cues.

There are software solutions to this. The screenshot in Figure 3.8 is from a typical ADR session, showing how takes are recorded and chosen by the ADR mixer.

There are also workarounds, depending on budget; one doesn't need to invest in hardware and software packages to produce workable ADR. Todd-AO has an app that lets talent work remotely – the ADR editor pre-loads the visual cues as clips, which are sent to the actor, who then views them on their phone or tablet. As they record their takes, they're recorded onto their device, and the new material is shipped back to the editor via the cloud, where it's re-synchronized and edited. This approach can be a little more time-consuming than it would be if all parties were together in the recording studio, but it came about as an efficient workaround during the pandemic. It's proven convenient for actors who end up on far-flung sets without a recording booth close by.

A lower-tech alternative is loading the film onto a phone or tablet so the talent can reread or simply parrot their lines after viewing them on their device. No booth necessary, and this method gives the sound designer the flexibility to better match the original environment where the line was recorded.

In addition to matching the performance, there is also a need to achieve a technical match with the original on-set dialogue. It is, therefore, vitally important to know what microphones were originally used on set. ADR editors will talk to the location mixer about this, but hopefully, the mixer also included all pertinent information in their sound logs. Using the same make and model microphones in the booth for the ADR recording process as were used on set will go a long way toward achieving the best match. Still, as mentioned before, mixers also need to consider the environment and how to best match that as well.

For this reason, most voice and ADR booths are treated to be as acoustically "dead" as possible – meaning only that, in addition to being isolated from the rest of the world, they are constructed to be as minimally reflective of sound within as possible. This lack of acoustics is preferred because adding reflections after the fact is simpler than trying to remove them after the fact. For instance, there would be almost no acoustical reflection in a scene with two characters conversing in an open field. Making even the most isolated interior space sound like a reflection-free exterior space can be challenging, so great care must be taken. Additional sound-dampening materials can be added to

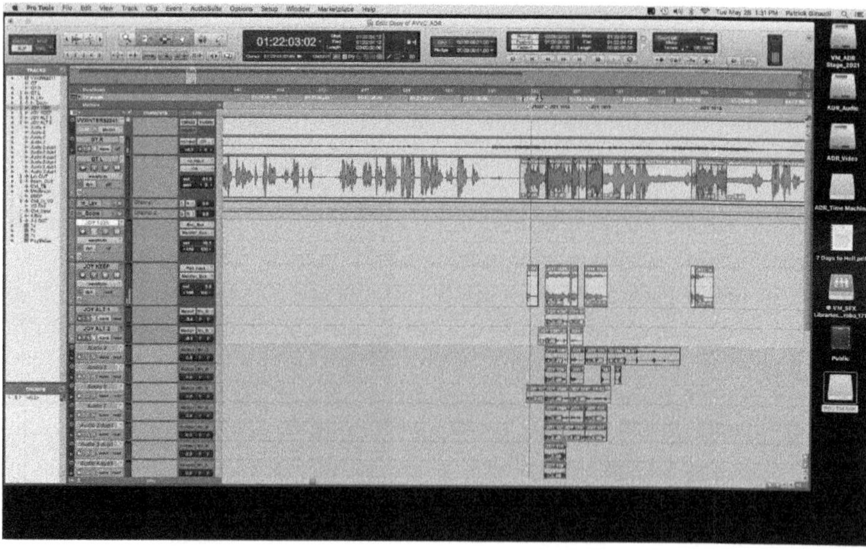

FIGURE 3.8 An ADR session from the engineer's perspective. Lines are recorded on the timeline, then the best take is selected by the dialogue editor.

Source: Permission H9 Films and Moody Street Pictures. Photo courtesy Mark McBryde.

help with that, but it can be a tough task. If those two characters in a field are raising their voices, they also need to match that projection level in their performance. Even the most isolated space will introduce some unwanted reflective properties as their projected voices mean a greater chance of hearing new reflections in the confines of the ADR booth.

Ideally, the ADR stage can contain those reflections and won't introduce any new, unwanted reverberation into the takes. Isolation like this is a requirement for any ADR stage, but even the best stages will have some issues at louder sound pressure levels; it's simple physics. Additional damping can sometimes be brought out to help with that, but there are limits to the efficacy and practicality of that, depending on the performance. One lower-tech possibility would be to take the actors back to that field to redo their lines, or at least an acoustically similar exterior space – bring a tablet for video and a recorder to capture the performances. Logistically, it's more complicated, but it would be more accurate sonically.

As the ADR and dialogue editors are doing their work, the sound effects editors are the next team to get their hands on the project. As with every other part of the filmmaking and storytelling process, there is usually a division of labor here, with tasks delegated and divided amongst the team.

Here, we encounter one of many seeming ironies in sound work – while going to great lengths to exclude incidental, non-dialogue sounds on set, the sound effects team will have to add them back in later. On the surface, this may seem counterintuitive, but there are many practical reasons for doing it this way, with the main motivation being the ability to maintain control over the quality of every sound, its level, and its placement.

Every sound is an opportunity to further the storytelling in some way, so keeping all non-dialogue sounds out of that track sets up better control of the character, placement, and quality of these sounds. But there's also an ease-of-use element to this, as keeping sound effects separate and distinct from dialogue makes it far easier to control and manipulate them, often in very different ways than the dialogue. It's also vital, in the end, to keep the sound effects tracks and stems separate for international delivery. All of this in the name of organization and control, where each sound can be given its due as part of the storytelling soundscape – and while, individually, those incidental sounds might not seem to be crucially important or central to the storytelling at any given time, they do need to be present as part of the sonic texture of the moment.

In one example, consider a dinner scene with normal-to-quiet dialogue, where our actors are actively eating, drinking, and serving themselves while delivering dialogue. The problem first presents itself in production when the mixer can hear the clinks and clatter of plates, glasses, and silverware happening simultaneously with lines of dialogue. Some of those sounds, being highly transient in nature, might begin to overtake the dialogue in level.

More experienced actors may know to time their movements so those actions and sounds happen separately from the lines being delivered, but it's not the responsibility of the actors to make this call. If the director isn't aware, the mixer is responsible for pointing it out (usually to the assistant director), to try to coach the actors to change their timings. Ultimately, it may just be that dialogue and a noise-making action will have to coincide at points and one must accept that Sound will have to do some ADR and sound design to recreate the moment.

However, it is always easier to head off these issues before they overtake the moment, so it is up to our production sound crew to do their best to quiet these sounds to whatever extent possible. They may opt to put down a coaster or a pad for glasses and cups to land on, helping mitigate those moments and actions. They will work with the Props and Set Decoration Departments to remove the ceramic plates and replace them with paper once the shots have gone into isos and those items are no longer visible – there's no reason for the noisy plates to be there if they're not on camera. If actors still need to appear to be eating, any food can be transferred onto quieter paper plates at this point instead. And, of course, they can try working with the actors to get them to time their lines better and in between noise-making actions.

This last point is politically fraught on some sets, however, and it is important to note that, on a larger production, this kind of communication with the talent often goes beyond the production mixer's responsibilities – they will need to go up the chain of command to get this accomplished. If the relationship exists, they could approach one of the assistant directors. The sound team can also try quietly interacting with other departments (Set Decoration, Props) to get some help with knocking down those noises. On smaller productions, the mixer should probably take that bull by the horns and do their best – but again, production is a team sport, and it's sometimes a short leap from taking the bull by the horns to actually becoming the bull. Bulls tend to trample the things that get in their way, and the last thing any of us wants is to undermine the authority of those further up the chain.

As with the earlier processes, the sound supervisor must outline the work they need the sound effects team to accomplish. They will work to figure out the scope and the scale of the work to be done and then assign it to the individuals best suited to execute.

It is important to note here that the supervisor is not a micromanager – although that might have been the example in the pre-digital era, it was sometimes necessary from an efficiency standpoint at that time. In those days, their time would have been better spent on the administration and organization of work to be done, i.e., laying out sound effects pull sheets and mapping their locations – than to have sound editors "on the clock," as it were, attempting to make their own best choices while combing through linear sound libraries with a lengthy and tedious process of auditioning and selecting sounds, and then dubbing them off, as they pursued their particular sound editing tasks.

The digital era flipped that model on its head, to an extent, and significantly streamlined the process of work of the sound effects editor. The editor now has a much faster workflow at their fingertips, allowing them to audition and manipulate sounds to a far greater extent than ever. As such, sound effects editors are being hired as much for their aesthetic sense and taste as for their technical prowess – we are all, in effect, sound designers these days, and the sounds an editor may select for an edit are chosen in a much more autonomous mode than in those earlier times, in a process that is also much less equipment-intensive. The digital revolution has democratized the process, and rather than an expensive flatbed editor in an even more costly post-production facility, all any sound editor needs these days is a moderately powered computer and some software, and this work can be done literally anywhere in the world.

As with the dialogue, a series of assessments will be made early to determine what sounds might be covered in the production audio. It's usually not very many, as the on-set crew will always be more focused on getting dialogue and excluding those sounds as much as possible. The sound supervisor will make their best determination about what sounds might be able to be sourced from existing sound effects libraries and what sounds need to be created out of thin air. This last set of bespoke sounds is broken out even further into more granular sound categories.

First, to further help the credibility and believability of a scene, there are incidental sounds, such as footsteps, the sounds of props being handled, and other objects being touched, handled, or manipulated. These sounds will all need to be recorded in a way that optimizes these sounds for the action being seen on screen. The incidental sound effects captured during production will likely be off-mic, as the focus of the recordist on set was to get clean dialogue, so all those actor-induced sounds will need to be produced and recorded in a way that feels realistic to the viewer. The level of detail requires the Foley team to get into the subtle details of clothing rustles, body movement, and hand sounds. All these sounds fall under their purview, and they will set up a space (the Foley stage; see Figure 3.9) to re-perform those sounds quickly and accurately, all in the name of furthering the scene's believability.

The Foley team's efforts are a requirement for any work that will see international distribution. The reasoning is that when synchronous dialogue is eliminated (as it would be for international release), there is also a loss of presence in some of those small, incidental sounds that make it feel realistic. Certainly, there are footsteps, but clothing rustle, hand sounds, and all the other sounds a body inadvertently makes in the course of simple action and communication will all need to be recorded after the fact. In the end, these will all need to be added to

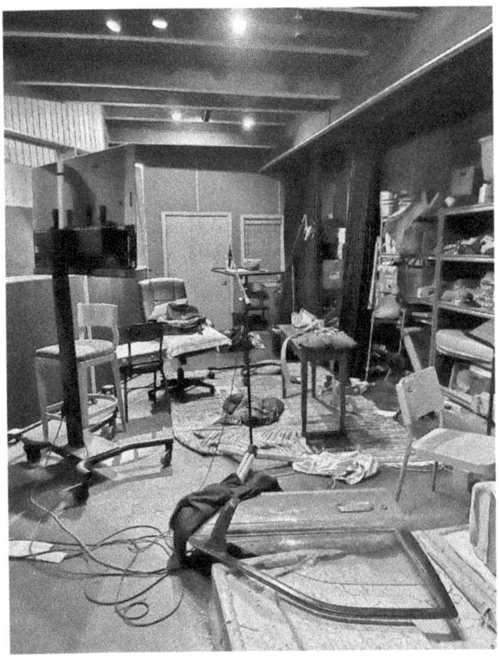

FIGURE 3.9 Foley stage with props

Source: Courtesy Mark McBryde.

or replaced so that they are present on that international set of deliverables and help further the believability of that version of the soundtrack.

It can be a lot of work for what seems like a small return, given the fine detail; it's certainly not as glamorous as planets exploding and spaceships flying by – but when done well, just that little bit of detail is what it takes to sell the viewer on the idea that all that action on screen happened in real-time. Returning to that video example of contextual sound, well-executed Foley functions as a lens, bringing the viewer's attention to a specific action as one player fidgets nervously. That sound helps point out what might otherwise have seemed an insignificant action; it was subtle enough that it might have gone unnoticed.

In another scenario, well-placed Foley effects can help set up a scene or a situation. Imagine a character making their entrance into a scene – the sound of their footsteps approaching, along with those other incidental sounds, can point out that they are being introduced as someone important, perhaps central to the story – the sound treatment points them out as a more significant part of the scene than just a background actor passing by. It's a subtle redirection of the viewer's attention with a sound that, most of the time, they might not even be aware is happening – they are being subtly told to look at and consider this new person as they arrive.

To create this crucial work, the Foley crew typically consists of at least one Foley artist – sometimes called a "walker." At this point, the craft is so much more than that, but the term has stayed for decades. The other critical team member is the mixer. In the context of the Foley team, the mixer is the person tasked with all things technical, including setting up the microphones, recording everything properly, and adjusting sync on the takes to prepare the Foley tracks for whoever on the sound effects crew might be dealing with them next. In this context, "mixer" seems a bit of a misnomer, as that person is the central engineer; they're not just mixing but also recording, editing, and generally running the technical side of things for any given Foley session.

All this work is done on a Foley stage, essentially a purpose-built studio for recording these incidental and prop-generated sounds. Refer to Chapter 2 for a more complete overview.

There are often dozens (maybe hundreds!) of small, subtle sounds to be recorded, so for this team to be the most efficient, they must also be very well-organized. The first step for the Foley team – sometimes in consultation with the sound supervisor – will be setting some parameters to determine what exactly is required of them. Many tasks are assumed to be in the purview of the Foley team – those feet, hands, and cloth sounds are generally a given – but creatively, someone needs to direct this small unit and make the call as to whether there are additional, special sounds the sound effects team might need to pay attention to.

The first task is to create a cue sheet, which can be seen in Figure 3.10 in a more analog mode.

SOUND EFFECTS COLLECTION LIST BY SCENE

Production Title:		Recording Engineer:	
Director/Producer:		Date:	

Scene | Sequence Description:		TC IN	TC OUT	Duration
Scene Number:				

Hard Sync Effects	CD	Foley	Wild	Soft Sync	CD	Foley	Wild	Ambience	CD	Foley	Wild

FIGURE 3.10 Foley cue sheet

Sounds are typically organized by type – the aforementioned feet, hands, and cloth, but also what is referred to as "sits," which covers any action where an actor is moving into or out of a sitting position on any type of surface. There are likely to be many other actions that need an accompanying sound, including anything and everything the characters may move or handle, as well as any special actions, sounds, or items. As one can see from the cue sheet, conditions might also need to be noted here. What kind of footsteps are being placed or replaced, what type of footwear is on screen, and what kinds of surfaces are they on? Are they interior or exterior? Wet or dry? A Foley artist will have any number of shoes and surfaces to work with on their stage, so any description that might help with the execution is important to pass on and will help streamline the planning process.

Take feet and footstep sounds as an example. A cue sheet is compiled for the first character seen on screen – call her Character 1 – and would include all instances of her footsteps in the film. It would also denote what the conditions were for all her "walks," including (but not limited to) whether she wore shoes or was barefoot, a short description of the footwear, what types of surfaces she is seen walking on, whether they were interior or exterior, wet or dry – and all these cues will be organized onto one cue sheet. This allows the Foley crew to skip to the scenes where she appears, set up for any special conditions, and record all her scenes while in that setup or mode. Working non-linearly like this dramatically speeds things up; it helps the Foley team run more efficiently and stay on task with each Foley grouping. The Foley crew will repeat this process as they perform the rest of the Foley work, helping them make the best use of each setup before moving on to the next.

The paper cue sheet mode of working is outdated but helpful as a fallback if needed. More often, the team will utilize a digital audio workstation for this process, creating tracks populated with blank clips to serve as the cues and labeling the tracks and regions with the work to be done (see Figure 3.11). Efficiency is paramount, and this mode allows the Foley mixer to load the multitrack workstation project into their system and simply follow each track to the next cue. All Character 1's feet sounds will be assigned to one track, Character 2 on another, clothing and cloth sounds on another, and so forth. The editor will end up with many more tracks this way but can quickly parse the work into more manageable sections – and ultimately, make quick work of the Foley shoot for the story.

Once finished, the mixer will select the best takes of each – having taken accurate and detailed notes along the way, of course – and package them in a cohesive, concise way for the next step. This prepared "Foley shoot"

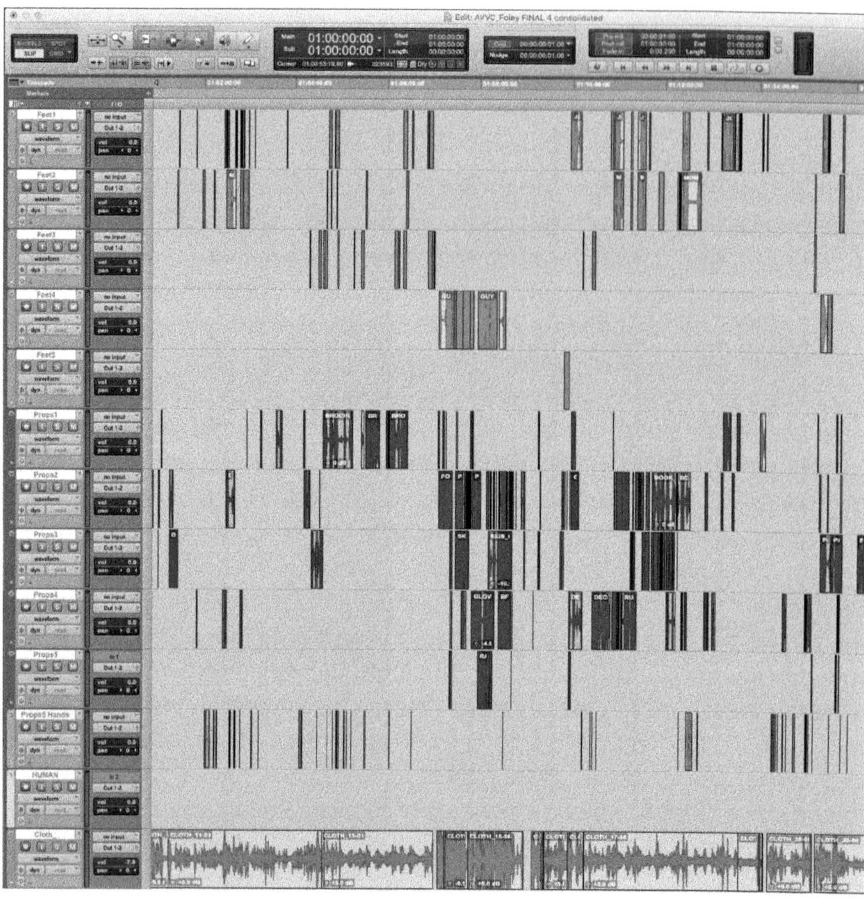

FIGURE 3.11 Foley cue in Pro Tools

Source: Courtesy Mark McBryde.

(sound-speak shorthand for the Foley session with all its files) will then be handed off to a sound effects editor, who will incorporate those tracks into the larger build for the mix sessions and predubbing later.

While the Foley crew does their work, the rest of the sound effects editorial team will work on the film's foundational sounds. Another fundamental truth in sound work is that the sound effect an editor needs is inevitably the one they don't have. Many pre-recorded and packaged libraries are widely available – Sound Ideas, Hollywood Edge, and Pro Sound Effects are some of the more popular pay-to-play libraries. There are also any number of crowd-sourced and free-use effects libraries available, so in the end, there are many ways to source the sounds needed to begin a sound design.

The main advantage of these databases is their ease of use – setting up to record sounds can be expensive and time-consuming. For certain staple sounds, it's simply a labor-intensive and inefficient means to an end. For more unique sounds, such as antique vehicles, certain locations, settings, and even some props, obtaining the access or items needed to record those sounds can be challenging.

Many of the more foundational sounds can be generic – think city sounds, room tones, and the like, or what are typically called soft effects, referring to the nature of synchronization these effects might require. Soft effects are generally descriptive sounds and have little to no sync basis – it doesn't matter how they're placed to picture; they have no sync points.

At the same time, there are also any number of generic hard effects, such as doors, light switches, and so forth, that are synchronous to a corresponding action on screen and are therefore considered contextual sounds. Sounds sourced from a library for these purposes are sometimes almost mundane, but they are necessary to help build a specific reality within the scene.

There are many available sound file management programs and applications to organize these sounds, such as SoundMiner, BaseHead, and Pro Sound Effects' SoundQ, to name a few. These platforms will organize and scan a sound effects database (and some will also look to cloud-based libraries). When a search is initiated, the search engine will present the editor with a list of sound candidates according to the input search criteria. From that list, the user can audition those sounds and place them directly on the workstation timeline, where needed. This is a far cry from looking through physical catalogs, as happened in the days of vinyl and, later, CD-based libraries, while also a considerable improvement over doing a filename search of a hard drive directory for that elusive effect. In addition to the auditioning and placement, these programs will help the editor with list management, tracking sound effects usage, and other reporting functions while allowing new sounds and libraries to be imported into an existing database as new sounds are produced or acquired.

When the production has specific sound needs that the libraries can't cover, the sound supervisor may need to set forth and go into the field to record new sounds that are entirely proprietary to the production. The soundtrack's bigger and more dramatic aspects, such as specific vehicles, environments and ambient sounds, machinery, and so forth, will require a more deliberate and bespoke approach. A special team of recordists will be assigned to go out into the world to stage and gather specialized sounds, coordinating the many sound sources and the locations needed to record them.

The sound effects crew can see their assignments become ever more granular and specific, depending on the size and nature of the film. For example, the typical Hollywood blockbuster might see several teams of sound effects editors, each brought in for their specialty and expertise. These teams can encompass many areas, but specialists can be found in the industry who bring their own experiences and (perhaps most importantly) their own bespoke collections of sound effects. For instance, if the film at hand is set in the Old West, that would set a few parameters for the sound design. Certainly, there will need to be horses, wagons, and other horse-drawn vehicles and all the sound and clatter that would accompany them. It would also mean an absence of certain sounds, such as the freeway near the set or the airport flight path right over the location, which might require even more sound work for the team.

Hiring an editor with a reputation for doing this kind of work would be the sound supervisor's first priority, as that editor would also likely have an extensive database of horse and wagon sounds already at their disposal. Taken together, it will also lighten the sound supervisor's workload, as less time will need to be spent directing them (in theory, at least), and fewer resources will need to be spent tracking down those horse-specific sounds.

One of the sound editor's greatest assets lies in their sound effects database. These sound collections, usually amassed and curated over many years, are likely to include some more familiar, widely available sound effects libraries, but editors and sound designers will take care to retain their bespoke, specially recorded, and proprietary sounds as well. Any use of the editor's database would be negotiated at the outset because it represents such a significant investment of time and expense for that editor. On the other hand, the sound team might only need a hired hand to round out its sound editing staff and require an editor to work only from their in-house, already bought-and-paid-for (and cleared) sound effects.

Whether a sound effects library has been purchased or personally acquired, it will be curated and managed by the sound effects editor who owns it. In most cases, sound effects are free to use once purchased – unlike music, there are not usually rights issues with the sounds purchased from a sound effects library clearing house. Editors will often be required to compile sound reports and generate a sound effects log – something quickly and easily done from the digital audio workstation, but as one would imagine, it can be a very long list.

Original sound effects are also often created for a specific production, as this approach is just the easier, most expeditious path. It can fall somewhere between Foley and sound effects work at times, but "shooting sound" is an important tool to have in the arsenal when the existing sound effects library falls short and there's no practical way to perform the sound required in a recording booth or Foley stage. This would include unique ambient sound from different locations or under odd or unusual conditions, the sounds of specific machinery or vehicles, or any other real-world sounds beyond the practical use of a Foley stage, such as weaponry, groups of people, animals, etc.

Sound designers like Mark Mangini travel with a handheld recorder on the off chance they will hear something unique that can be banked away for a future project. As the sound professional's work takes them to various locations and requires ever more diverse sounds, it helps to be prepared for any unusual sound events. Curating a personal collection to suit one's immediate and future needs means capturing everything from oddball vehicles, machinery, nature sounds, and ambiences, to weather events, animals, and crowds – the list is endless, and one never knows what might be needed in the future.

3.2.4. Coordinating with the Music Department

The musical aspect of a story can contribute significantly to the perception of the story as a whole – music can provide an emotional context and do it in more direct ways than sound design can. At other times, music can underscore the pretext in a moment or foreshadow events by leading us to the next.

A composer will come on board to create original music cues, or a music supervisor can help map out the musical treatment in a film. Their comprehensive knowledge of pre-recorded music can lend a different musical context to the film through the use of familiar music, sometimes by showcasing current artists' newest works.

Whatever the path to that final music track takes, it always comes down to the Sound Effects, Dialogue, and Music Departments working together to create that greater whole in the form of the finished soundtrack.

The sound supervisor will certainly interface and communicate with the music composer or music supervisor, but the sound super's authority is by the mandate of the directorial team. While it's always helpful to have cooperation and coordination between the Sound and Music Departments, the sound team doesn't have any say in what music is chosen for the film – and that mandate goes both ways. The best path forward is always one of cohesive, coordinated goals.

Ultimately, once again, communication and organization are the keys to success – having open channels of communication during the process will help prevent potential confusion and uncoordinated tracks on the mix stage later, euphemistically called train wrecks. Work timelines and the timing and execution of that work can differ between team music and team sound design – the composer might have been engaged earlier in the process, while, the sound effects team might be working right up until the mix begins. This effectively keeps both departments in separate silos.

Complicating things further is the definition of sound design itself. Some music programs teach sound design as a musical concept, and many music synthesis tools facilitate the creation of music-based sound design, so it can get confusing when a music person starts talking about sound design to a sound effects editor. Their two definitions of that term might have less in common than one would think.

Here is where that potential train wreck effect comes into play – no one wants to end up on the mix stage with the sound design interfering with the music or vice versa. For a soundtrack to work, it needs to function as a whole, with no surprises for the mixers on the mix stage.

The best collaborations come about when the two teams have open communication and can work out potential conflicts well ahead of time. As just one example, the sound design team might have an ambient treatment that has a tonal quality to it – it would be best to reach out to the music side to determine if that element will conflict with the music happening at that moment, or if there even is music happening there. Conversely, if the composer is thinking of incorporating sounds into their track that might veer into sound effects territory, they would need to check in with sound editorial to ensure their score doesn't double up on something the sound effects crew is working on.

The best composer collaborations happen when the composer can provide something – anything! – for the sound effects team to work with. Most importantly, this temp track shows where the music cues will be happening but also indicates the most important elements like musical key, arrangement, orchestration, and so forth so that the sound effects side knows what will work best with the music track at any given moment. It's important to determine whose team gets to carry the moment ahead of time – that way, they can better work together.

They can also narrow the choices the director will have to make in the final mix and prevent them from having to referee the two teams to determine whose work gets to carry the moment. Mark Mangini says there should be no surprises on the mix stage. While working on *Dune*, the back-and-forth with Denis Villeneuve had already gone on to such an extent that, by the time they got to the final mix, "Denis referred to the tracks and sounds as old friends, he already knew what was in there and how it was going to play." In this way, the work of the final mix could be entirely focused on making those tracks work together; there were no surprises.

Of course, in this "be prepared for anything" world of sound, it's important to both consider and allow for the fact that things might change. It can be difficult when planning on having either music or sound effects carry the moment in a scene, only to have the consensus on the mix stage suddenly change, and now the other element needs to carry the scene instead. If music is stripped out of a scene at this late date, the sound design for that scene must be complete enough to work in the absence of music.

It's important to try to avoid those moments, but in the rushed world of getting a story told, the tendency is to streamline the process as much as possible to avoid unnecessary work. Doing the work of scoring sound effects into a scene in the event they *might* be needed could rightly be considered a waste of time if they never see the light of day. This brings it all back to communication being the key to a streamlined workflow and helping head off those moments before the mix work begins.

3.3. ENVISIONING SOUND

A question often asked is where the overall vision and approach to the sound design aesthetic comes from.

The easy answer is … everyone. Visual storytelling is a group exercise; it's at its best when it's a team enterprise, with every member of the team there to bring value and perspective to the project. A director may have some very specific ideas in mind, but to be effective, that director needs to surround themselves with like-minded creatives who can help realize that story. It follows, then, that the director also needs to be open to the contributions of their team, understanding that the perspectives and creative approaches of others might not be exactly in sync with their own. Allowing oneself to be open to those inputs and opportunities is the hallmark of a great collaborator, setting the stage for creative experiments and happy accidents.

As Mark Mangini has stated, "I get hired for my opinion, and the director wants that opinion."

Directors and other project leaders may not have all the answers or have the vision fully realized at the outset, but letting creative people be their creative best is how a project can truly take off.

3.3.1. How It All Comes Together

Many assume that the sound and music teams don't begin their work until the picture is locked, but bringing on key leaders much earlier in the process can be a major benefit. A sound supervisor can make suggestions to streamline the production process and even help avoid potential sound issues if they see the script in time. At the same time, the composer might be able to supply some temp tracks for the editorial crew to work with in their pacing of the piece, and so forth. These early feedbacks and suggestions can inform later creative decisions, as well as heading off potential problems.

3.3.2. Delegating Tasks

The sound supervisor is closest to the pictorial team – primarily, the director and the editing crew. At the same time, they answer to the producers because a project must run on time and within its budget.

The sound supervisor – or sound super – will then build their team, hiring crew according to their strengths and experience. A dialogue editor, for example, might get the call because of their work in documentary films; their résumé would indicate they're effective at making real people sound believable and articulate. Another dialogue editor might be known for their work with action films, where a lot of dialogue must be replaced due to technical issues – pyrotechnics, stunts, and fight scenes all interfere with the ability to get good, clean dialogue, so replacement is all but assured in many cases.

From there, the sound super will bring on the others – Foley artists and their mixers, various sound effects editors, and so on. Again, all are being brought in according to their strengths and experience. Like actors, film tradespeople often find themselves typecast by their work, so for many, the gig being offered is going to be based on past successes. We need to remember that we're only ever as good as our last job. If an editor gains some renown for their last sci-fi film, chances are good they'll be offered another. But it's also a team sport, as said before – that team wants a team player, and how well we work in a group is a large part of landing the next job.

Tasks will be split up according to individual strengths and abilities, with the sound supervisor being the one who delegates those tasks, trying to get the best from the team while also meeting any time (and budget) constraints.

The sound supervisor is also the vital conduit to the director, from whom all creative direction flows. The director might not have any answers as to *how* a particular goal will be reached; that's the sound super's job – managing expectations and informing the director of what's possible.

There will be several screenings and demonstrations along the way – think of them as screen tests for sound, as a proof of performance exercise to show the director how it all works together, and as a vehicle for them to sign off on critical elements of the project so that time and resources can be delegated elsewhere. With the director's (and likely producer's) input, the sound design and editorial process will continue, ultimately landing at the mix stage with a complete track.

While this is all going on, the composer and the music crew will also be working their magic on the piece. There may be several people to delegate to, including orchestrators, arrangers, contractors (individuals who hire the musicians needed to perform on the music recording sessions), and the technical personnel in the control room – music recording engineers, mixers, and editors.

By interfacing early on with the music team, the sound supervisor works to head off any conflicts at the mix stage. The hope is that both parties will have a strategy session with the director before any work is done, to avoid conflicts and overlaps. This level of coordination can be a complicated dance at times, but it's important to have a roadmap in place so that all the tracks arrive, without conflicts, on the mix stage. One example would be a complicated action sequence, where the music score is working well and carrying the day with all the emotion and impact one would hope – yet, at the same time, the sound design side went above and beyond. Their treatment of the scene is also impactful. Which path should the director choose?

Truthfully, directors want to avoid playing the role of mediator or referee between the Music and Sound Departments, so there needs to be constant, high-level communication and interplay between the two teams as the soundtrack evolves and develops. The mix stage is not the place to have these battles; it's simply too expensive to make decisions then that could have easily been made in advance. There's also a lot to be said in favor of preserving the forward momentum of a positive, creative experience while there. This part of the process should be productive, with less labor and more love.

When it is done, the tracks all arrive at the mix stage for prep, premixing (or predubbing), and mixing, ideally without any clashes or conflicts.

3.3.3. The Mix Stage

In the analog media years, mixes would be prepped for the mix stage through predubbing. It was a process of combining tracks of similar elements into composites of those sounds, which would then be incorporated into the tracks used in the final mix. It was an effective tool to make the best use of a limited amount of media resources and hardware, as those tracks occupied space on physical media that had to be played back from multitrack tape machines or mag dubbers.

For example, assume a sound design for a busy office scene uses five layers in the ambience build. There will be a room tone, a voice crowd sound, different phones ringing, and keyboard clatter, perhaps with some footsteps and paper rustle. That ambient soundtrack would be taking up valuable space on five machines, so the team would re-record those elements onto one track – effectively freeing up four machines while making some early mix decisions in the process. This early subsection of the mix, part of the predubbing process, takes all the ambient sounds on that scene and combines the layers into one composite ambience track.

That process would be repeated on the effects tracks, combining similar elements, like footsteps and Foleys, hard effects, door slams, car crashes, gunshots, and other "soft" effects, such as cars or planes passing. How those tracks are mixed – or predubbed – was up to the sound supervisor and would vary by project, with the overall concept being to lower the track count to something more manageable.

In the present day, digital tracks exist in virtual space and are no longer constrained by physical media. Consequently, track counts have inevitably gone up, but so has our ability to deal with them effectively while keeping them organized. The digital predubbing process has fundamentally changed the current practice of premixing, where tracks are mixed in sections and families as before – ambiences, hard effects, and so forth – but the tracks can remain virtually separate elements. Workstations can allow the mixer to group tracks together in whatever configuration they'd like but then control that group with a single fader control. The tracks within that group are mixed relative to each other, which is the beginning of the premixing process, but here in the digital domain, it's done without the commitment that physical media requires.

In the end, premixing is an organizational tool to address the creative and aesthetic aspects of the mix while also managing the technical needs and logistics of handling what could be a lot of audio media in the end. In a large-scale narrative feature, there could be literally hundreds of tracks, all edited and prepped by the team. It would be difficult – to say the least – to arrive at the mix stage without at least some organization of all those elements. Figure 3.12 shows Formosa Hollywood's Stage 1, one example of where that final mix happens.

The final mix is where the magic begins; that final mix is almost a celebration of the culmination of all the department's hard work. The film comes together in its final form, fully realizing all those hours into one successful mix. Either way, this is when, if everyone has done their job well, the final soundtrack will bring that visual story into its final form, making it something greater than just the sum of the efforts that have gone into it.

The sound supervisor will sit in and may even participate in mixing the film. Representing the Music Department, the composer will also sit in during the mix, in part to make sure their contribution is well-represented in the film and clarify their intentions with the music score if that input is needed.

Each group of sounds will be mixed by a specialized mixer – usually one for each of the sound macros of dialogue, music, and sound effects. They'll constantly communicate during the mix to ensure everything comes together exactly right. This process can take days, sometimes even weeks, depending on the complexity and creative demands of the film. To an outsider, it can feel like an endless repetition of the same elements, effects, and even individual lines of dialogue, as they're played repeatedly as the mixers strive to find the exact balance.

Of course, this is the Hollywood, big-feature model, but in any mode, the final mix is a distinct stage of the process, with all the other work for the project culminating in this moment. All that work can make or break a film; it doesn't matter if it's the Hollywood, big film mode of working, with an aircraft carrier–sized console with multiple operators (Figure 3.12), a single mixer worksurface

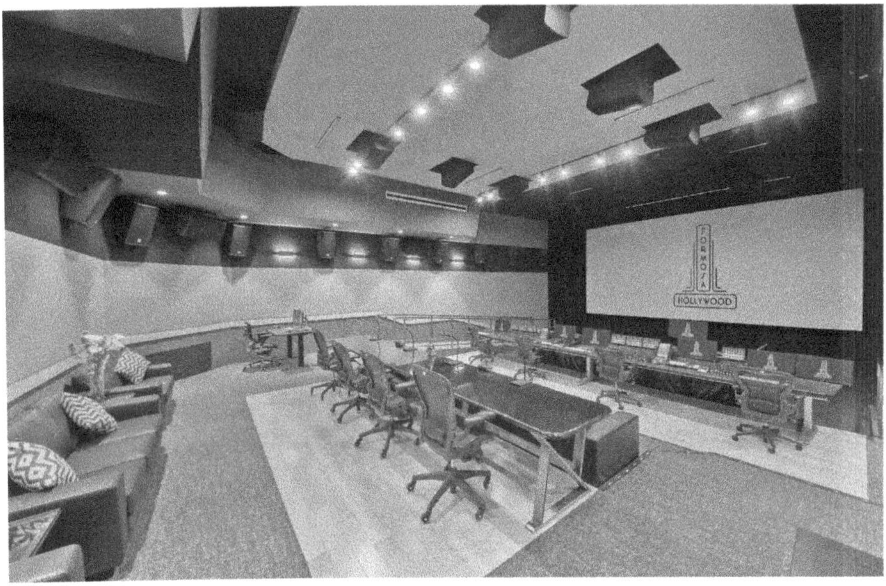

FIGURE 3.12 Formosa Hollywood Stage 1

Source: Formosa Group.

with multiple faders (Figures 3.13, 3.14), or even virtual faders on a laptop (Figure 3.15). The interface may vary, but the process is the same.

In all cases, knowing the medium – or outlet – the piece will be distributed on is essential. It used to be simple and obvious for film – big films play in big theaters, with big screens and big sound systems. Terrestrial broadcast television was a different yet predictable world with its own restrictive specifications and requirements.

The entire paradigm has shifted, and in more significant ways than anticipated. In this new, post-streaming, on-demand world of media distribution, films might initially be released in theaters, but they inevitably see release on any number of streaming outlets – sometimes concurrent with the theatrical release or after a short initial run in theaters. Television, as we used to know it, has shifted from the restrictive specifications of NTSC broadcast to HD, and programming can be seen in one of two modes: streaming or broadcast. The ratio has shifted in favor of the streaming side – the on-demand aspect of streamed content is far more convenient than the strict broadcasting schedules of terrestrial television. And then there is an entire universe of other online content, such as YouTube, privately owned websites, and so forth, that have yet other different technical specification requirements – if they even have one at all.

The perennial issue here is that it is almost impossible to accommodate all the ways a viewer might consume their media with just one mix. To that end,

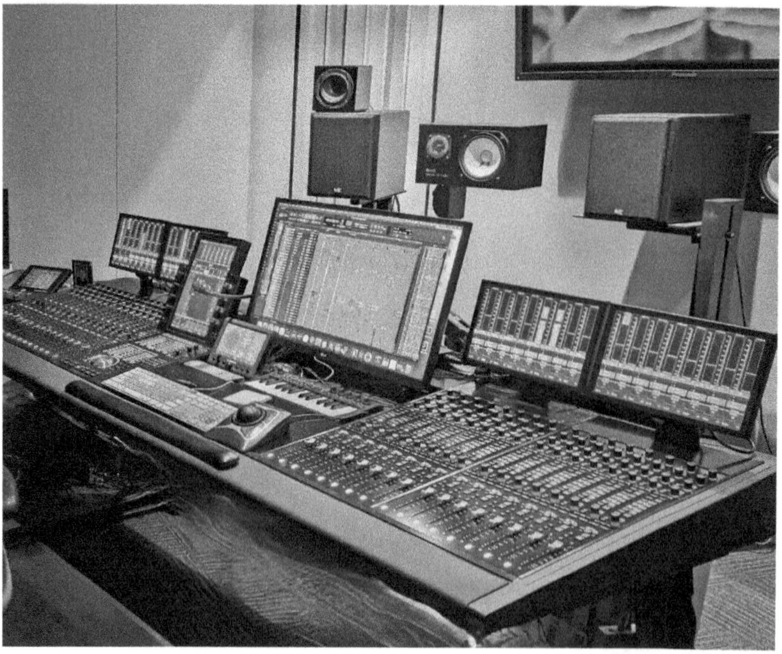

FIGURE 3.13 WGBH A Mix with S6 console

Source: Courtesy WGBH Educational Foundation.

surround and two-channel stereo remain somewhat constant, but many consumers still don't have surround-capable systems. So, it's safe to assume the majority of the viewing public experience the mix in stereo. However, the reality goes deeper, and a sizable segment of the viewing public gets their media experience in ways that are less than optimal for great sound. Phones, laptops, and tablets are all getting better-quality sound output than they used to, but the fact remains that they are still small devices that aren't designed to deliver the optimal sound experience in the first place.

Theatrical film aside, modern-day sound professionals are generally mixing for some version of the home screen television, with the handheld and portable devices being a secondary concern. A mixer should take time to check their work on a variety of playback platforms, cycling through different speaker output options to help ensure that the mix will "play" out in the real world and in a manner that is representative of at least the intent of the storytellers and content creators.

There are many sound workflows to help ensure a more cohesive sound mix overall. After all, if a mix sounds good on a small set of speakers while, at the same time, it sounds impressive on a larger set of studio monitors, then the mixer can be assured they have a mix that will successfully convey their intent out there in the world.

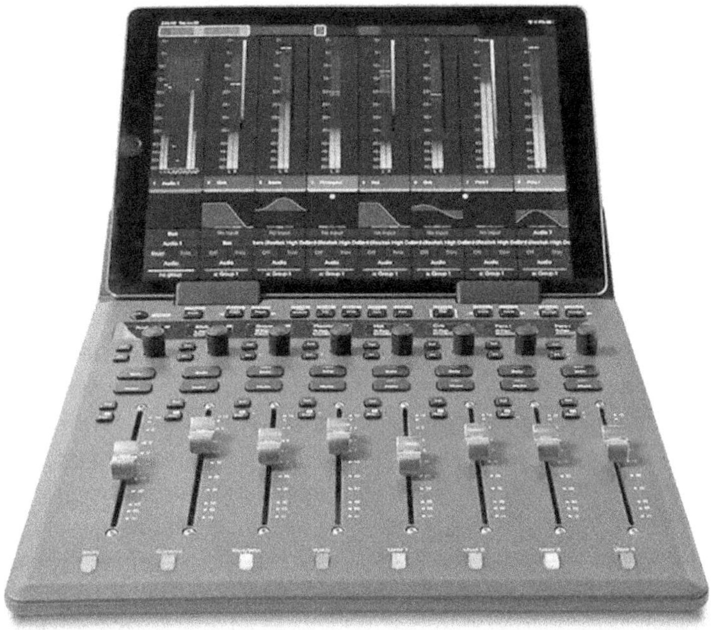

FIGURE 3.14 Avid S1 console

Source: Permission Avid Corporation.

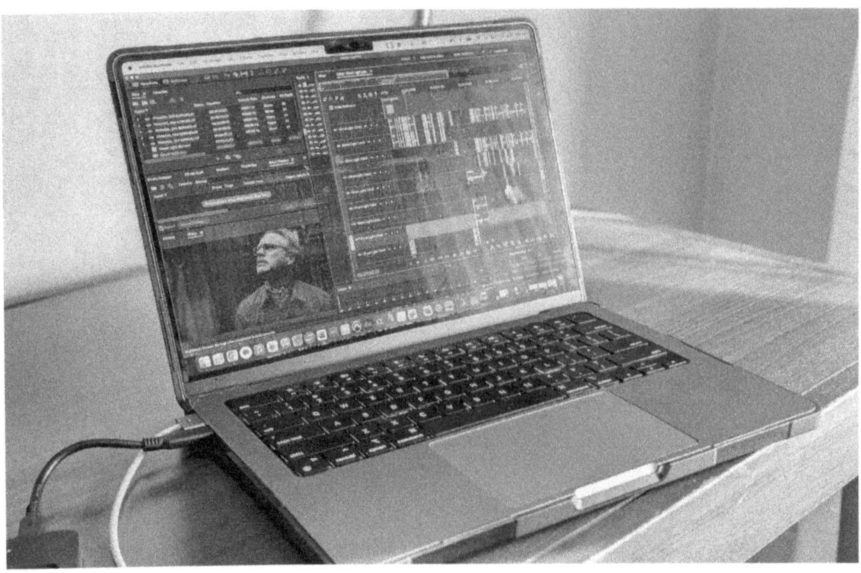

FIGURE 3.15 Adobe Audition on a Macbook Pro laptop gives us the portability to work on our tracks from anywhere

There are judgement calls to be made, for sure. A great example of this can be found in documentary filmmaking, where every syllable has weight and is essential to the message being put forth. The viewer can't be left behind or distracted by background or other sounds; the exposition of the whole piece is paramount. Contrast this with narrative filmmaking and some of the liberties a director might take with the sound effects and music being a more significant part of the story than they might be allowed to be in a documentary.

A great example of this can be found in the works of the director Christopher Nolan. In his film *Interstellar* (Figure 3.16), there are moments where the dialogue is overwhelmed by the rest of the sound treatment – music and sound effects both, at times, partially obscure the dialogue. At the time of its release, some critics were harsh in their opinions of that approach, but it's important to note that no vital plot points or conversation were obscured in those moments. Ultimately, the viewers weren't missing any vital information. Nolan, with his sound team, Richard King and Mark Weingarten, took the approach that those voices, at those moments, were meant to be as textural as the other sound design elements in the soundtrack. In its way, this is a more realistic approach, as it's unrealistic in the real world to always hear every single syllable of every word spoken. This fresh approach to treating dialogue as alternately narrative and sound design element is an important development – sometimes it's good to throw convention aside and try something new. At other times, it's essential to hear all the words. As the sound designer, it's your choice – but as with *Interstellar*, it will be up to the audience to decide if it's effective or not.

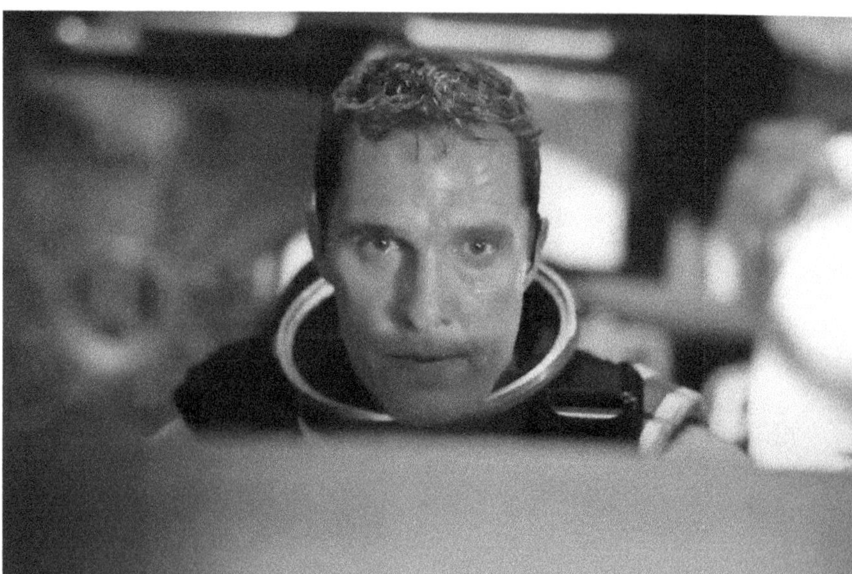

FIGURE 3.16 *INTERSTELLAR*

CHAPTER FOUR

Distribution – Getting It Out There

4.1. MIXING AND MASTERING FOR THE MEDIUM

The mix is where it all comes together, and all that hard work in sound editing, sound designing, Foley, dialogue replacement, and organization sets the stage for the mixing process.

As outlined in the previous chapter, knowledge of the medium is essential. The capabilities of all the potential outlet streams can vary, sometimes significantly. For example, a feature film would be mixed in a completely different style and manner than an online series. At the same time, linear media like terrestrial television shows may need to be handled differently than either of those. It's up to the mixer to take steps to ensure the mix will "play" – that is, it will sound clear and clean with nothing obscured – in the venue or medium for which it's intended.

In the world of music production, there is an entire post-mixdown process known as mastering. This is the very last stage in a very long process before the music is released to the public, where the mix done in the studio is further refined and optimized for the medium of release. This step occurs after the final mixdown in the recording studio and before releasing it for distribution.

Mastering is a way to optimize the sound mix so it will sound its best in all the possible distribution paths it might take when it has been released to the world commercially. While the music score in a soundtrack will continue to grow, evolve, and be both creatively and technically manipulated, the bottom line is that getting that music to work in the context of the greater film mix involves setting that track up to accommodate and work with the rest of the sound design and dialogue in that soundtrack. It needs to be strong enough to stand up to the full sound design treatment, yet also at times to provide some

DOI: 10.4324/9781003373094-5

space, allowing room for the all-important dialogue track, sometimes becoming subtle enough to appear on the verge of being barely audible, even subliminal – while somehow covering all the sonic ground in between. A music or song release, on its own, will likely not have to compete with the elements of dialogue and sound effects; it will probably be front and center during its time in the film. The challenge is to make that song full and compelling and as close to perfect as possible. In other words, perfect.

The needs of a music score are far different than they would be for a stand-alone music release. A score needs to work with other elements, whereas the song release needs to sound as full and fleshed-out as possible to stand up to the medium and the competition of the other music it might be grouped into.

Music mastering engineers are highly sought after for their technical expertise and discerning hearing abilities. They can implement the processes needed to make the music track work in every circumstance and over every possible distribution channel while also allowing for all the environments and modalities in which potential listeners will access that music. Mastering engineers are at the forefront of technology, but they are also potentially tastemakers to some extent, and the specialization within that community of mastering experts becomes even more granular, with different methods, styles, and tastes coming into play with each genre of music. For instance, when working with a popular artist, one would want to hire a mastering engineer with expertise in the popular music arena. In contrast, if the music style is in either the classical or the jazz domain, one would want a mastering engineer with strengths in those areas, as the tastes and sensibilities of the pop-style mastering engineer might not be the best choice for those styles of music.

In nearly every case, the music will be produced, recorded, mixed, and effectively locked in and done by the time it reaches the mastering engineer. The mastering engineer will then take the piece and work with an array of signal processing devices to optimize it for the many places it might go, including (but of course not limited to) broadcast radio, one of many streaming services, and home release, including CD, MP3, or WAV file format, or sometimes even vinyl or cutting-edge ATMOS or ambisonics. All these streams have different capacities, capabilities, and needs. With all the needs of these formats, there would be some differences in how a mastering engineer will approach them. Some things can be planned for – for example, knowing that the typical broadcast chain will have a dynamic limiter inline before it gets to the transmitter. That processing can sometimes wreak havoc with a mix that hasn't been mastered and pretested for that eventuality. CD and vinyl releases have a far better dynamic range than the broadcast chain allows, accommodating louder peaks and quieter valleys in the final program material, but very often, these are also the mix versions that will be broadcast, so they need to be set up for the effects the broadcast chain will have on them.

There are very different mastering requirements for the possible distribution streams, and vinyl, in particular, requires some additional and fairly aggressive processing treatment to get it to sound as good as a CD or file release. Then there are the streaming services, such as Apple, Pandora, and Spotify, where the mastering engineer needs to anticipate and deal with the effects of data compression loss.

In this make-or-break scenario, the mastering engineer needs to have the proverbial "golden ear"; they know the technology well enough to arrive at a successful end, combined with an aesthetic and musical sensitivity that can further ensure that a given song will stand up to the various processes, and still sound as good as its peers.

The reason for bringing this all up is that an equivalent position doesn't exist in sound for visual media, so it falls to the mixing engineer, or re-recording engineer, to ensure that the soundtrack holds up to all the modes of public consumption it will need to flow through. This often means creating multiple mix deliverables, where different versions of the mix are designed to accommodate all possible media formats and platforms. In every case, it is entirely up to the mixer to ensure that the sound of the story will be robust enough to represent that soundtrack accurately in all its potential distribution channels.

4.1.1. On Mixing

There are many ways to approach a mix, and although it may seem the most mysterious of all our sound processes, it's within reach and a skill that can be learned over time – with some trial and error and accumulated wisdom. It might not be an answer the aspiring mixer wants to hear, but no one formula or algorithm will make a better mix. It's an acquired skill, developed over time and with experience, with intuition and "feel" being part of the calculus determining exactly what makes a good mix. The mixing process is both technical and creative, and it requires the engagement of both halves of the brain simultaneously and forcing them to communicate.

"When talking about mixes, we're including the premixing phase. On a big film, we will premix dialogue for about two weeks, then we premix sound effects for about three weeks," says Mark Mangini. This is the last round of preparation for the mix before everyone arrives at the mix stage to finalize those elements into a coherent final mix.

"Then we're in the final mix for about two weeks. Once the final mix has been accomplished, everyone will give notes; then we will go back and update again, do more notes, and then we're done, and we head to making the print master."

Mixing is about managing the interactions between sounds and getting them to work together in a way that supports the visual story. It begins in the sound design and editorial process, where the vision is first realized and comes to life as tracks are edited, adjusted, and combined to find their place in the overall track build.

Sound is ephemeral, and one cannot simply freeze a frame of sound to examine it. The true medium is time, regardless of the media – tape, film, or hard drive. Humans can only experience sound linearly, and it can't be broken down into fractions of a second to scrutinize, manipulate, and enhance.

Knowing the final mode of distribution is critical and essential information to have when developing a coherent mix; the capabilities of the format will dictate, to some extent, the creative and technical approach to the sound design in the first place. For example, it might not make sense to design a broad, dynamic, cinematic track in full surround format, with all its bass and glory, for a piece that will be distributed only via podcast. Similarly, a cinematic feature requires a much deeper sound field than a broadcast outlet can handle; the two are worlds apart. Certainly, it's good to have options available in the track build when mixing, as it can be hard to predict how it might all come together. It's always better to have a track with more options than might be needed over a build that, in the mix session, turns out to have a few key elements missing. It's also essential to understand the medium to know what sound design will work in that soundtrack in the first place.

It's also important to understand the objectives of the mix before beginning. Knowing what the distribution channels are is critical, but ahead of that – what is the purpose and intent of the mix? Is it a nonfictional exposition of facts or a narrative work? Is it likely to be seen and heard in the theater or the home? How will it be delivered – via individualized channels, such as streaming services, other online outlets, or terrestrial broadcast channels? In documentary works, the dialogue and narration elements are usually given the greatest priority; they're the most crucial element of the exposition of the material, there to present the information in a logical, consumable way. At the other end of the distribution spectrum, an action film will require a bigger-than-life sound design treatment to go with the visuals; big sounds and a dynamic music track help to relay that sense of tension and excitement.

It's easy to assume that greater track counts mean better sound, but that can result in a lot of extra work for the sound designer. Letting the story dictate what is needed in its telling can make for a cleaner, more accessible mix instead of falling into the "more is better" school of thought. Having some options in the final mix can be helpful, but the truth is that overbuilt tracks will always result in a lot of unnecessary work for the mixer – it's essential also to remain selective when choosing sound elements; sometimes, taking the more straightforward path will be more effective. So many sound designers engage in what we call the "pasta treatment" – throwing every element at the wall to see what sticks. Just as with cooking pasta, the only result is a mess for someone to clean up – in this case, the people on the mix stage.

When mixing, taking an organized, deliberate approach can be helpful, bringing in elements by groups. Identify the element most important to telling the story, a process that usually starts with the dialogue element. Utilize equalization, compression, and other signal processing to ensure the voice has the clarity and presence to stand up to the rest of the tracks that will follow. It can be hard to gauge in advance, but these are also impermanent decisions that are simple to revise later – at this point, it's a matter of taking some educated guesses as to what might work best. This is also the time to begin setting up reverbs, futzes, and other specialized processing to place dialogue in a particular space.

Mark Mangini provides some context from the world of feature film mixing, talking about his work on *Dune,* where they premixed dialogue for about two weeks, then worked on mixing sound effects for an additional three weeks. Premixing and predubbing are essentially the same processes in modern workflows when mix personnel assemble and process the various sound elements to prepare for that all-important final mix. The difference is that, in the past, premixing meant committing to that premix and committing it to physical media through the process of recording or *predubbing* it to either tape or magnetic film. This was done to conserve media – track builds could easily exceed the number of physical machines available, so sounds were combined into groups to create composite elements in a series of these "mini mixes." Subsequent predubs would then be combined with other predubs to get the track count to a manageable place that matched the number of machines and physical tracks available.

In the current digital audio architecture, predubbing is the same process it was in the analog mixing era; it is essentially a means of organization and making those complicated mix decisions in smaller, more manageable steps. The tracks still get mixed in the same way in the end, the main difference being that they remain separate – and therefore adjustable – throughout the mixing process.

Much time and consideration go into how those tracks are represented in the final mix. After a lot of listening, reviewing, adjusting, and undoubtedly no small amount of second-guessing, the elements are deemed ready for the final mix stage. Mangini recounts, "On *Dune*, we final mixed for about two weeks, and then we had a number of review and tweak days, taking another three or four days to play back." The final mix could only happen after those five weeks of mix preparation to bring the tracks to a place where all parties felt ready.

It should be noted that the dialogue mix also precedes all the other elements in Mangini's world of feature films. When the voices are edited and processed and seem to fit into the context of the soundtrack, that element is secured, and all the other tracks – music and sound effects – can be mixed to support that main narrative element.

Equally crucial to the process is the time it takes to properly review and understand how the mix is working as a whole. It can be challenging to objectively review a mix when so much time is spent working on it in smaller increments of time; it can be like trying to view an elephant with a microscope – we're only examining smaller parts of it at a time, without the continuity and context to see the bigger picture. Until we get that final review and play through, it has been a series of segments to the mix team, with each one essentially under that microscope. The context of the work only comes into view when the mixing team can step back and review it in the fullness of its runtime to experience how it all plays and works together.

It also helps to gain some perspective with some time off between the mix and those review sessions. Getting enough rest is important to the process as well.

4.1.2. Evaluation and Specifications

As if managing the creative and aesthetic aspects of the sound design wasn't enough, strict technical requirements are also imposed on the mix. The creative treatment must conform to the requirements of the distributor's deliverables.

Most outlet and distribution streams have a set of specifications that all the deliverables must adhere to, with broadcast and streaming services being among the most stringent. They do this to meet a whole host of technical parameters but also to provide a consistent listening experience. Historically, this has not always been the case – the limitations of the media were the common denominator that largely dictated any standardization of the audio deliverables. So first, a little history.

4.1.2.1. The CALM Act, Level Specs, and How We Got to This Place

In addition to its picture limitations, analog television imposes some serious limits on sound. Headroom, noise issues, and standardization across delivery mediums were constant issues that had to be accommodated and managed – and often, simply lived with. The broadcast world has seen a lot of evolution and growth in the last 30 years as we've transitioned away from mono mixes, analog media, videotape, and the standard-definition television signal standard to high-definition television and all-digital broadcast signals. This upgrade in quality has meant new, updated workflows as the industry has adapted to entirely new technologies. Growing pains in this drawn-out adoption process were inevitable.

On the audio side, working in the digital domain has provided a lot more headroom and a much more comprehensive dynamic range to work with. The evolution was haphazard as broadcasters moved forward in fits and starts as their broadcast chains were upgraded. However, once the industry got there, and especially as digital broadcast became the standard, the genie was out of the bottle, and sound designers and mixers found they had a much broader dynamic range to work with.

As a result, television programming content had a lot more headroom to play with; with the lack of consensus on a standard of sound level, the levels of the content often varied significantly from program to program, as well as with the commercials in between. Having some programs seem louder than others provided an inconsistent experience, to say the least, and was proving to be annoying and distracting to viewers. As viewers complained, broadcasters began to specify certain audio operating levels to help maintain a more uniform listening experience between programming for their viewers.

The wild card in all this came from the commercials, which were often not as scrutinized and were largely unregulated. Because they were coming from different distribution sources than the programming – and, it should be noted, were paying a high rate to keep that programming on the air – they were mainly being given a free pass when it came to their loudness levels. It was a bit of a loophole, and mixers who specialized in mixing television commercials started mixing those commercials to be just as "hot" or as loud as they could technically be. And again, the viewers complained.

While it would have been difficult enough to go between programs that were all mixed at different levels, it was utterly disorienting and disconcerting to be watching a program and then have a commercial show up as a kind of sonic assault; they could be startlingly loud in comparison to the adjacent programming, upsetting the viewer in the process. After a barrage of complaints in the United States, the limitations of the Federal Communications Commission (or FCC) were revealed to be ineffective and without any mechanism in place to meaningfully address the situation, Congress were prompted to get involved.

After a series of hearings, Congress passed the Commercial Advertisement Loudness Mitigation Act, more commonly known as the *CALM Act*, in September 2010, following the recommendations from the Advanced Television Systems Committee. The *ATSC* is an international body whose guidance has helped provide framework and technical specifications for the broadcast television industry over the years; they helped develop the guidelines for Congress to include in their legislation. President Obama signed it into law in December 2010, empowering the FCC to implement the new law. The new legislation was enacted in December 2012 in its final, slightly revised form.

Historically, the major broadcast networks were largely self-regulating their content's sound levels up to this point. Still, again, commercials were often outside the control of various outlets, as they would be sourced from many different places. Advertisers wield a lot of power in this dynamic, and every ad client would want their commercials to be just as loud as possible – at least as loud as the competition and the surrounding ads a viewer might see in the same ad break. The problem was widespread, as the same commercial might be seen across several outlets as part of an ad campaign.

Complicating the issue further, a local broadcast affiliate could be airing a nationally broadcast program while, at the same time, holding some of its commercial slots aside for local advertisers, amongst the other placements already claimed by the national advertisers. The national commercials would often be programmed in place by the broadcast originator, essentially being a part of the programming that local affiliates re-broadcast, giving them no direct control over that element. They could only manage the local advertising they had sold and placed, but even then, they had to heed the demands of their local advertisers and allow the local commercials to be just as loud as the national spots. The worry was that otherwise, the local commercials wouldn't have the same impact if they weren't as loud; they would be somehow diminished by being quieter and might not be taken as seriously. This set the stage for some real loudness wars in commercial advertising through the late '90s and into the early 2000s.

A big part of the problem was that post-production mixers were very effective at their craft and were intentionally using their tools and skills to make those spots louder; they knew how to optimize their processing gear for maximum effect. Along with effective sound design, they employed multi-stage, multiband compressors and limiters. They applied them in sometimes heavy-handed ways, helping them get a commercial to register average sound levels at the very top of the measurement scale. Because of the lack of restrictions and oversight, mixers could often do whatever it took in their quest to make their ad mix the loudest spot on the air.

As mentioned, networks were already working very hard to keep their programming sound levels consistent. Because programming might be coming in from a variety of production house sources, the networks needed a way to ensure consistent levels for all those submissions. To help with all that, Dolby ultimately provided a technical solution in a metering system that could measure *average* level and peak information in one device – so the networks put this to work.

The CALM Act gave some of the control back to the broadcasters, making them ultimately responsible for the loudness of the commercials while at the same time controlling their network content. They could then adopt those same types of level standardizations for commercials and get them into a place consistent with the programming specs the networks already had on their own program material. The spec has evolved and been refined over the years; the latest iteration is the ATSC A/85:2013 measurement for most broadcast outlets.

The spec calls for an average level, measured over the program duration, of –24 dBFS (±2 dB) – meaning, the mix level should average out to measure at –24 dB on a full-scale digital meter when calculated for the total duration of the programming it is being applied to, give or take 2 dB either way. This measurement method is different from the real-time metering used in many systems. It measures an average level over the length of the program, so it takes some time to get used to it operationally.

As one might guess, getting a program to land at precisely –24 dBFS average is difficult, so networks and broadcasters allow for a variance, usually ±2 dB on that specification. This means the mix level average can be as high as –22 dBFS or as low as –26 dBFS.

As we know, a digital meter goes up to 0, where anything above becomes distorted. To provide a level of protection from extreme sound level peaks and possibly constraints in the broadcast chain, the spec also calls for a True Peak measurement of –2 dB, or 2 decibels below that 0 mark for momentary peaks (see Figure 4.1). Because those peaks are generally transitory and very short in duration, they have little impact on the overall average level measurement.

It can be challenging to adapt to this mode of mixing, as the meter measures an *average* level over time, meaning that level readings are drawn from

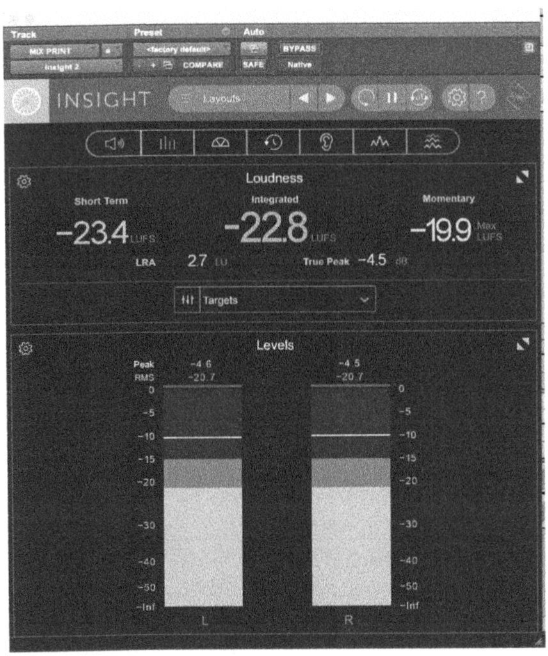

FIGURE 4.1 Izoptope Insight 2 level metering plug-in

Source: Permission Izotope, Inc.

the entire length of the program material. It is not a real-time meter, so if it shows that a mix is out of spec, it's up to the person mixing the program to figure out where in the timeline the violation occurred – depending on the length of the program, there could be several spots that are slightly too loud, bringing the average up, whereas in shorter programming (such as a 30-second television commercial) the entire programming might need to have its level brought down. It's very important to know the window of time being measured – the specification is applied to the duration of the entire program, not the snapshot in time that a traditional meter provides.

It can be an issue for those of us working in post-production, especially when considering that the entire program isn't viewed in its entirety until the end of the mixing process. To get around this, meters can be set to a "short term" mode, where they show the average level over the previous 10 seconds, while some meters utilize a histogram type of display to help provide the operator a more accurate representation of where any peaks might have occurred. This helps make tracking down and managing those out-of-spec moments easier.

The takeaway is that a level measurement of a 30-second TV commercial would be calculated from its short duration. At the same time, a 53-minute documentary would see its average computed over its entire 53-minute duration, giving it a larger set of level data to draw from. Commercial television gets even more complicated, with the measurement often applied to the program segments between the commercial breaks, each measured individually.

It has certainly added a complication to audio post-production workflows, but at the same time, it has had a smoothing effect over the years, helping each network's broadcast space sound and feel more consistent overall. The unintended consequence is that commercials can still feel significantly louder than the surrounding programming they're plugged into. Instead of aggressive mixing techniques, it's now a function of the relatively short window of time being measured and the dynamics of the program itself.

For example, a television drama that airs terrestrially on one of the main broadcast outlets in the United States will likely have quieter and louder moments. The quieter times help set up the louder moments and help build drama, being designed into the mix from the outset. The quieter moments also offset the louder ones, helping the program average to meet specifications, so all is good.

The discrepancy arises when a 30-second commercial is inserted into that program – it still must deliver its message and make its point very quickly, so its overall loudness is set as close to its maximum average loudness as possible; it can be mixed to stay right at –22 dBFS throughout its duration. Technically, that's legal (±2 dBFS allows it to be as high as –22 dBFS), but it feels like a wall of sound coming after the up-and-down dynamics of the programming it was inserted into. The average loudness of the program material could be very different from the loudness measured at any given moment, as it has a broader range between its dynamic peaks and valleys. The spec accommodates this tradeoff, giving the sound team the creative space to design a more exciting and dynamic program.

While the commercial will seem louder, it technically is not; its loudness is simply at a consistent level. This can be jarring for the viewer, coming out of that dramatic program with its up-and-down dynamic – we could be going from a quieter moment in the program to a commercial, with that commercial mixed at the maximum level allowable under the broadcast specification.

This is the nature of things now – the program requires some dynamics, or it would be hard to listen to. If it were consistently at the same loudness, viewers would tune in to something else, as it would be tedious to listen to for very long. The commercial needs to make its point in a much shorter window of time, so it must take advantage of all the headroom available to it.

4.2. FORMATTING AND DELIVERABLES

Loudness control and management is just one of many challenges facing the mix team; they also need to be at least cognizant of how many different potential modes of viewing their media will dictate how the program needs to be mixed.

How can the final track be presented in a way that sounds good on all those distribution channels downstream from the mixing stage? Like everything else in the media arts, there can be a lot of different answers, and it often depends on who is asking.

Many mix rooms have different monitoring modes available to the mixer at any time. The concept is that the mixer can switch between modes during a mix to hear how that mix might sound on different monitors, or speakers. Different monitors serve other purposes in the mixing environment, with the big, expensive speakers mounted on the front wall (Figure 4.2) being the first step toward getting a clean and clear soundtrack. These are the speakers we see in a high-end recording or mix room; they're sometimes almost comically oversized, they always sound amazing, and they are sonically accurate – and so, they can reveal many truths about the sound going through them.

Sound designers and mixers strive for accuracy in their work but sometimes struggle with how the equipment between them and the audio media might be coloring the sound. Mixers and editors need a highly accurate speaker system that does not add anything to or subtract anything from the sound and lets them hear all the nuances – and potential issues – within that sound. We call this quality *transparency*, where the speaker is only the vessel, passing the sound through to us as accurately as possible.

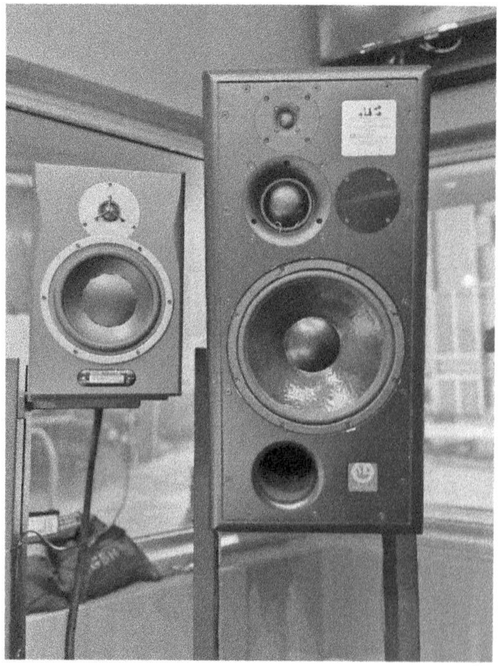

FIGURE 4.2 Focal speakers

Source: Permission WGBH Educational Foundation.

While those big, beautiful speakers are impressive, expensive, and sound very good, there is a point in the mixing process where they can actually sound *too* good for what we're trying to accomplish. That quality of transparency shows individual sounds and combined sound elements in a completely unvarnished, honest way. However, they also sound so good that they don't always reveal the relationships between the different sound elements within the soundtrack. Trying to judge where, in sonic space, the music track needs to be in relation to the dialogue can get tricky when the speaker makes everything sound good. We can accurately hear the individual sound elements, but it can be difficult sometimes to determine their relationships with each other. Is the music too hot? Sounds great! Is the music too soft? We can't tell, but it still sounds great!

To counter this, mix rooms usually have a pair (or two) of smaller-format or nearfield monitors. These also sound very good, certainly better than many consumer-grade speakers, but they are slightly less transparent, making them more helpful in revealing those soundtrack relationships. Less transparency means more sound coloring, with the qualities of the speakers coming into play. Different speakers serve different purposes in revealing the relationships

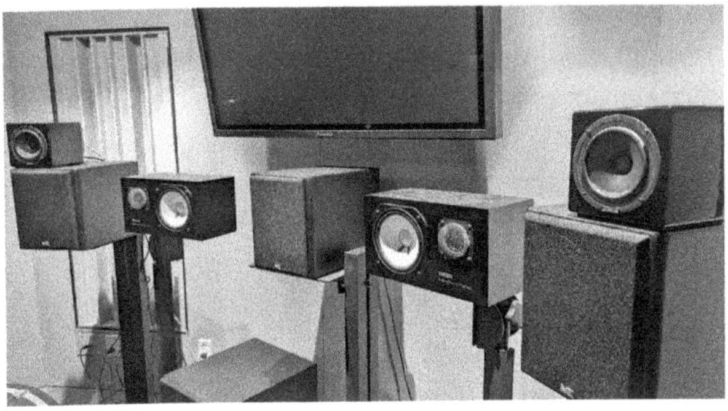

FIGURE 4.3 A control room with a variety of monitoring options. Avantone mix cubes are shown on the far left and right, more closely representing the listening experience many viewers may have.

Source: Permission WGBH Educational Foundation.

between the tracks in the mix, and audio engineers gravitate toward their favorites.

At the far end of the speaker spectrum, many studios often have speaker options that are closer to consumer-grade, coloring sound in ways that trend closer to the average home viewing setup. These speakers might not have the incredible bass response or highly defined upper frequencies of bigger, more expensive speakers, but most home listening environments do not have those, either. These are the "common denominator" speakers, like those in Figure 4.3, that the mixing team might use as a last check of the mix, playing it through something closer to what the average viewer would have in their home.

In addition to speaker options, some professional studios might have alternative methods for checking their output. Some music production studios have used low-powered FM radio transmitters as a double-check; after a mix pass, everyone can go to their car and tune in to hear the mix and experience it over a system they're already familiar with. If a mix is going out as a podcast, it would make sense to check it over a smartphone speaker or on earbuds, as those are two of the most likely means of consumption for that kind of programming.

It's important to point out that headphones aren't always the best choice for making critical sound decisions, particularly when it comes to complicated, multitrack mixes. Like those big studio speakers, professional-grade headphones are also typically very transparent, making judging the distance and relationships between dialogue and the other sound elements difficult, at best. If it's necessary to mix on headphones, then it's essential to do a speaker check later, just as a double-check to ensure the headphone version of the mix is a good one.

In the end, the choice of playback monitor speaker system should be one that is familiar and trusted by the mixer. The reality is that many viewers will likely not have access to the same quality speakers or listening environment as a recording studio, so the only way to compensate for that is to listen to the mix on as many different playback systems as possible. Porting a mix out to a familiar device, whether it's a television, portable Bluetooth speaker, earbuds, or a laptop, is an essential step to help the mixer better assess the strengths and weaknesses of the mix and then adjust accordingly.

4.2.1. Distribution Formats

Beyond making it sound good, there are many ways to finish that soundtrack and send it out into the world, with mix format being one of the most significant factors in the equation.

In the early days of film with monaural sound, there was only one mixed track to worry about: the monaural format, referred to as simply "mono." Mono mixes remain a deliverable requirement; they still have an application as the basis for the Second Audio Program and Assistive Listening tracks (SAP). In this mode, the viewer can choose alternate languages or a narrated descriptive track, where a narrator describes the actions and settings on screen for the visually impaired.

From there, the possibilities grow into multiple multichannel formats that help make the soundtrack more immersive and realistic. Stereo formats – technically defined as two *or more* channels of sound – grew from the idea of creative, more realistic, and immersive sound fields that could trick the audience into believing they were in the world of the film, if even just for a minute. After that, there are several and various flavors of "surround sound," which take the listener deeper into immersive territory by literally surrounding them with multiple speakers playing several channels of sound. Sound design grew and developed this concept of the "sound field," and there is no going back.

Immersive sound is a great innovation, and as more sound channels are pointed at the audience, the greater the potential for not just realistic sound fields, but even more hyperrealistic, or even surrealistic sound fields and effects. No one has ever heard a spaceship flying through space at the speed of light, as one example, for a whole host of reasons – not the least of which are the practical realities of the physics in that statement. Humankind can't (yet) travel at the speed of light, no matter what all the best sci-fi movies say, and even if we could, there's no sound in space – so the list of improbabilities goes on. Still, as a fantasy experiment, it is a wide-open creative opportunity for the sound designer to come up with what that sound could be, how it might work in an immersive sound environment, and, best of all, how it might function as a means to further the storytelling narrative in new ways that no one has ever heard before.

But the question remains – what is the best way to deliver that experience to an audience? First, looking at the moviegoing public and the theater venues where the film would play is a necessity. Typically, theater venues offer a superior filmgoing experience, as they have better visuals, higher-end sound systems, and the means to deliver that experience more viscerally than any other venue. However, the public that consumes this kind of media doesn't always go to the theater venue to consume it. It can be impractical for some, due to the expense of going out, possibly hiring childcare, the cost of gas or carfare to and from the theater, and so on – to say nothing of the time consideration. The fact that most of this media soon becomes available for streaming at home continues to cut into theater revenues.

The simple reality now is that broadcast and streaming television remain the primary and preferred means of delivery for most of the public, but each comes with its own needs, capabilities, and demands. Real-time terrestrial broadcast continues to fall behind streaming outlets in popularity, but it is also because of the on-demand promise that streaming delivers to viewers. Watching a program on one's own time is so much easier for most episodic and dramatic programming, as it accommodates the viewer's schedule – instead of the other way around. At the same time, real-time broadcasting remains the primary delivery mechanism for timely, live events, such as sports, news, and weather.

Streaming here generally refers to the mainstream content providers, such as Amazon Prime, Netflix, Apple, and so forth, but for many reasons, these outlets should not be confused with YouTube. The primary difference for anyone in the media production business is the level of technical control over the content. The subscription streamers have production and quality control measures in place, whereas YouTube remains an outlet for crowdsourced material without real oversight or very much in the way of technical requirements or specifications. To be fair, this is not necessarily a bad thing – it's just a difference in delivery and expectations. With the paywalled streamers, one can expect programming across that platform to be of a high and consistent level, production values will have been given greater consideration, and there is an assumed level of quality inherent to justify that rental charge or monthly subscription fee.

Crowdsourced content, while democratic and largely unfettered by the whims of oversight by a larger corporation, undoubtedly will show some gaps in terms of content quality. And while there is room for some great content on the platform, there is a lot of content with some egregious technical issues. For our purposes, the streamers are at least curating the content and maintaining quality by ensuring that it meets some level of technical proficiency and specification via their submission standards, whereas YouTube does not.

4.2.2. Broadcast, Streaming

As said before, broadcast, or terrestrial television, has lost significant market share to streaming services, but live television still has some advantages in the arena of live events. However, for most narrative, documentary, and episodic viewing forms, broadcast and streaming take essentially the same process path through the mix suite, technical specification-wise. They can handle stereo and multichannel (5.1) sound formats, and both require the same quality control over their content. Some streaming services have been slower to adopt level specification targets as part of the delivery requirement – see earlier references to the *CALM Act* in the United States – but these specifications deserve strong consideration when producing material for any kind of distribution. Some streaming service providers are only beginning to realize that standardizing delivery specifications is a great idea, particularly as the platform might already curate thousands of titles. No viewer wants to be constantly adjusting their volume up or down to hear their favorite programs' soundtracks comfortably.

In terms of multichannel delivery, mixers will typically deliver an 8-channel mixdown, in what is referred to as a "5.1+2" audio master. This term refers to the two versions of the mixdown contained in the same delivery system. The first part, 5.1, refers to the surround mix (Left Front, Center, Right Front, Left Surround, Right Surround, and LFE, or subwoofer), which has become standard practice when delivering any program for broadcast or streaming. The "+2" refers to the stereo version of the same mix, which is often created by a fold-down algorithm applied to the surround mix, combining the six channels into one stereo output. This track is provided concurrently with the surround mix.

There is no situation where a viewer would ever hear both mixes at the same time, which is a good thing, as the summing of the two would be significantly louder than spec and likely out of balance within itself – the stereo mix would reinforce the Front Left and Front Right of the surround mix, making the sound effects and music seem quieter than they were intended. The decision of which mix to listen to is up to the viewer, or that decision might be made by the equipment available on their media viewing setup, including the soundbar, the receiver, or the video monitor itself – as well as how that sound signal is being delivered into the home.

In addition to the final mixdown, there is a whole schedule of items to be delivered to the distributor – these are called mix stems, also referred to as splits. Industry standards typically call for separate dialogue, music, and effects (DME) stems derived from the final mix. Sometimes, the list will include a music-and-effects-only (M&E) version of the mix – essentially the same mix but without the dialogue tracks, so that international distributors can replace it with translated dialogue for the market. Documentary productions will sometimes ask for a DME deliverable when there is narration present, representing the mix without the voiceover track, making it easier to dub in the language for the market where the piece is being distributed.

Further adding to the list of deliverable requirements may be what are referred to as "undipped" deliverables. This would be the same mix but without the dips in the track that would work around narration or sometimes dialogue. This is a particularly important element for documentary distribution because translated narration doesn't always have the same timing the initial narration might have had. The corresponding fader moves, or "dips" in the track, could very well end up in the wrong places, so it helps to facilitate the mix on the other end to have the track as smoothed out as possible, with those dips flattened out.

The schedule of deliverables serves several purposes for the distributor. First, it fills an archival role by giving them all the materials they might need to recreate the mix at any time. There could be several reasons for this, such as new information that has come to light (in documentary works) to alternate versions of a film for different outlets, such as broadcast or commercial television, or airline versions (usually censored and edited somewhat from the original), to music rights issues, and tailoring it to fit important international markets.

In all instances, it is essential to have the distributor's specifications on hand, as some requirements might vary, and then adhere to them exactly. Many distributors will go to extra lengths and may only accept a program that has been properly evaluated by a reputable, third-party house that does this sort of work. Technicolor is one of the giants in the industry for this service, but there are many others, and they all are there to provide an objective, third-party evaluation of your program to ensure it adheres to all known industry and aesthetic standards. They'll evaluate the program, with all aspects of video, color, audio, and layout, under the microscope, and then provide a detailed report at the end. That report will outline items that should be addressed and repaired, all graded on a scale that goes from highest priority – that is, items that are deficient and must be dealt with – down to specific items that could be a judgement call between direction, production, and technical requirements. Sometimes artistic choices can be a factor, but those might not always be immediately obvious to the technician evaluating the program. In some instances, the creative team is allowed to contest some of the lower-rated items on the list, but it can become a question of choosing the battles, as too many flags on a program might result in the distributor rejecting it entirely.

4.2.3. Film

Film sound distribution needs are very similar to broadcast distribution in some ways but with a looser level requirement. Broadcast and streaming media are generally mixed to a narrower dynamic window than film, while a film's soundtrack can take better advantage of the dynamic range available to maximize the dramatic impact. It all flows from the idea that in cinema, our audience is captive within the confines of the controlled environment of the theater, which has been designed to minimize distraction – outside factors that might compete for the attention of the audience have been reduced or eliminated. The quieter moments can be truly quieter, making the viewer focus more on them when they happen – which will cause the louder moments to be even more dramatic when they occur.

"When we start a mix, I'll draw a loudness map before getting started," says Mark Mangini. "We can better plan for the peaks if we know where the valleys are, and we can also create a build throughout the film, culminating with those bigger moments being truly big enough to have the impact we need them to."

Film venues also use bigger projection and speaker systems, which require a slightly different approach to the processing in a mix. The LFE, or subwoofer element, is just one example of how the theater environment differs from the home. There are certainly home sound systems that can deliver a cinematic sound experience, but the theater is better suited for this, with high-end sound equipment and better acoustic treatment and sound handling capabilities.

4.2.4. Podcast and Streaming Platforms

Here is where standards become more divided. Some platforms – YouTube, private websites, TikTok, and other social media – don't impose or adhere to any meaningful audio level specifications. In all cases, it's essential to ask the important questions earlier in the process and ensure the mixdown will work for the platform at hand.

For many streamers with a terrestrial broadcast presence, the specification remains consistent across both their broadcast and streaming outlets. This way, there is some portability in the programming, and programming is already cleared for either (or both) the broadcast or streaming channels. Streamers who maintain a cable tier presence, like HBO, Lifetime, Cinemax, and Showtime, will apply the same spec to their cable broadcast and streaming content. The objective is to unify the experience for viewers, whether they're tuning in to the cable broadcast or the on-demand version of the programming from a streaming platform.

Other outlets that are typically streaming only, such as Netflix, Hulu, and Amazon Prime, will follow the same standards because they want to maintain the continuity of their viewers' experience across all outlets – but also, much of their content might be coming from some of those cable and terrestrial sources as well. Many streamers are repurposing their content, re-streaming shows and programs previously syndicated in broadcast, so they want their proprietary programming to better match those older series that appeared previously on television. It simply won't do to have different shows at completely different mix levels, and it would be confusing – and maybe even a little stressful – for the viewer to have their experience upended like that.

4.2.5. VR and Gaming

Some see virtual reality as the next frontier in the viewing experience. Imagine being enveloped by the story's environment, both visually and sonically. As of this writing, we aren't yet at a point where that scenario becomes the standard mode of receiving media content. However, it's important to note that there are facets of this experience that could very well translate to current programming and, certainly, vice versa.

The idea of the viewer being enveloped and immersed in a virtual field of sound and visuals is very new territory – in fact, it's revolutionary. Questions remain as to the viability of different platforms and how this material might be delivered uniformly, as there is currently no specification or standard at this point in the evolution of the technology. Many of its developers and proponents are still very much in the development stage, and all are still vying for technical supremacy to help dictate and control the future standard. Apple, Meta, Google, and others are all actively working on their versions of VR, and it may be a while before the dust settles and a clear winner emerges.

4.3. SUMMING IT UP (PUN INTENDED)

Mixing is too often viewed as a dark art, and when the technical requirements of a mix intrude on the creative component – and vice versa – confusion can ensue.

There are a lot of tools in the mixer's arsenal, from the equipment and software needed to the techniques, tricks, hacks, and workarounds we all learn over the course of our experience. It might help to look at it as a skill to be acquired, much as one might learn to play an instrument or train for an athletic endeavor. The best practice is, in fact, practicing and refining one's approach, listening to the results in as many modalities as possible, and taking that experience to refine one's approach further again.

There are no shortcuts, although, while the latest technology and AI haven't yet come up with an actual "automix" capability, that time might be nearer than we think. Some tools help with the mix, but in the end it will always come down to human sensibility and aesthetics as the final arbiters of what a good mix sounds like.

CHAPTER FIVE

That's a Wrap – an Afterword

The more I know, the more I find there is for me to know. And I will happily admit that I wouldn't have it any other way.

It might seem odd to quote a master of the horror genre here, but Stephen King once talked about how we each have our own mind filter; what gets caught in one person's filter is going to be completely different than what shows up in another's. What he meant was finding a proclivity toward doing a certain thing or activity that leads us to seek out certain pursuits in life. He talked about how his own filter is designed with little horror story–shaped holes in it to let that stuff through so he could further mine it and refine it into the prolific body of work he is known for today.

In telling stories in media, we all have filters that help shape our proclivities and sensitivities, but maybe with little visual story–shaped slots in them instead. Or, to stretch the analogy even further, they could be sound design–shaped stories, as is the case for me. It could be an aptitude, maybe it's a calling, or maybe it's just a favorable bias – or maybe it's just semantics, I don't know. Call it what you will, it's been one of the few constants in life that I enjoy and I like to think I've gotten pretty OK at it, to the point where I seek it out, always wanting more.

I am a process person; I always want to try to find the most expeditious and effective path toward the most effective end result. Fortunately for me, there are many paths to a satisfying end result, but unfortunately, there are many more paths to that result that I find I need to sort through in my efforts to determine which one might ultimately be the best.

It is my hope that this text opens the door to more questions, more research, and a greater propensity for experimentation and pushing those boundaries on your part. Without sounding too "back in my day" here, there have been

DOI: 10.4324/9781003373094-6

some momentous changes to the technology and methods we use to go about our daily work, and the field of sound design is only getting better with every new iteration. Of course, these same kinds of advances have occurred across all the crafts and processes, not just in sound – and yet, despite the amazing advances in the various technologies and workflows that resulted from them, the basic principles and concepts of visual storytelling have remained essentially the same. We still must conceive, write, produce, and capture all that media, and then edit it into a concise and effective story; those core values remain intact. It's the more flexible technological processes that have afforded a much greater level of creativity than ever before – with greater speed and higher fidelity.

As I've said before, a sound engineer or film editor from the 1950s would probably be awestruck at what our technologies can do here in the 21st century, but they would at the same time innately understand our processes and workflows, once they got a handle on it. The awestruck aspect would come into play when they realize the speed with which we can get there. What might have taken that person – or even a team of people – several days, weeks, or even months (if it was even possible at all) can now be done in hours, sometimes even minutes.

The real constant here, just as in life, is change. Although it was an objective to try to "future-proof" this text to whatever extent possible, I realize that it too will inevitably become dated in due time, all because change is relentless. Texts written only 10 or 15 years ago didn't have anything to say about streaming channels of distribution because that technology was only in its infancy at the time; it was not always reliable and still largely unproven. Similarly, I don't have a lot of insights to offer here about artificial intelligence, as it's still very much in its infancy at the time of this writing. There will undoubtedly be new means of distribution and media consumption that will change the context of media texts like this one. There will be changes in processes and technology, from sound gathering to noise reduction and audio repair, that will steadily advance the idea of what is serviceable sound, while AI will allow us to create entire tracks from nothing but a small sampling of original material – or maybe even nothing at all, guided only by our creative input and aspirations. We're already seeing so-called AI-driven plug-ins, along with speech synthesizers and AI visual engines that can de-age actors, even bringing deceased personalities back to virtual life.

Along with these advances will come technical, ethical, and financial implications. It seems harmless enough that we can de-age an actor for a key flashback scene, but it won't stop there. As an example – just because we *can* bring Humphrey Bogart back to life in a film, in both speech and likeness, we need to also address the question of whether we even *should*. Who gets the benefit from that virtual performance, both creatively and financially? Who might be harmed by that new performance? Parties to the cause could include Mr. Bogart's estate, his old studio, or other, newer business partners that he never even met. Do Mr. Bogart's heirs and estate stand to reap any kind of reward – or are they even due any recognition at all? As we're just now becoming aware of some of these issues, it's important to know there will be many more to come, with only some of those outcomes known, and many more still unforeseen. Once again, technology has outrun the ethics of the situation, which is far behind the pace of the advances.

In the very early days of digital audio workstations coming onto the market, I remember a demonstration that a sales engineer did where they re-edited former president Richard Nixon's famous resignation speech. While the editing wasn't anything we couldn't have previously achieved on analog tape, it was the speed of the editing process itself that was most impressive, with the ability to audition many options, to instantly undo them, and the workstation's ability to better mask our edits with crossfading and other techniques. The result was an effective re-edit of his famous "I Am Not a Crook" refrain, wherein they had Mr. Nixon proclaiming the opposite, saying "I am a crook," among many other things.

On the face of it, this would seem to be nothing more than just a harmless exercise executed among self-proclaimed audio nerds for their own amusement, but it is also an early example of how the public's faith in media could be undermined. It was, in fact, an inexorable first step toward the "fake news" cries of today, when facts don't always align with expectations, while at the same time pointing out the responsibilities we in media have when it comes to the proper representation of our source material.

As it turns out, it's only too easy to mislead and misdirect, and while basic parlor trick misdirection is all fine and good when watching narrative, fictional film – even expected – any misdirection is in complete opposition to the mission when working with news and documentary material. The viewing and listening public must be able to have faith in our intentions and in how we handle the source material as it's presented for viewer consumption.

All to say, we need to be responsible out there – and just because we can do something, it's important to look deeper, and see if all the reasons also point to whether we should. It's one thing to create the sound of a spaceship that doesn't exist anywhere but in the VFX department's servers, but it's completely another to fabricate the words of a world leader or policymaker.

This is not to say that we shouldn't be pushing boundaries and trying new things while pushing the bounds of the latest technology – in fact, quite the opposite. The point is to make tech work for us, and in service to the story, while also somehow avoiding the pitfalls of that equation turning itself around. It's fine to say, "I want to show *this* kind of story, how does the technology serve us in that mission?" but not OK to say, "We have this kind of technology, now what kind of story do we want it to tell us?"

CHAPTER SIX

A Semi-Complete Glossary of Terms

6.1. (SOME) TERMS YOU'LL WANT TO KNOW

Many terms in the worlds of sound and media arts will be helpful to know, especially when dealing with a more experienced sound person. However, this glossary is by no means complete; its purpose is simply to provide you with a baseline of the term knowledge that consistently pops up in the field.

5.1 – One of several surround formats to be considered for distribution. 5.1 is the most ubiquitous, appearing in both home and theatrical situations, but there are many other surround mix formats – 7.1, 6.1, SDDS, ATMOS. The mix format will dictate some of the best practices in post-production, so it's essential to know the outlet requirements and final distribution format from the start.

5.1+2 – A deliverable requirement, noting that the 5.1 surround mix (see above) includes a stereo (LtRt) **fold-down** of the mix.

Ableton – A digital audio workstation software package commonly used in music production and post-production sound.

Adobe Audition – Adobe Corporation's digital audio software. One significant benefit is its compatibility within the Adobe ecosystem; Adobe Premiere projects can open directly in Audition and vice versa.

ADR (or **Automated Dialogue Replacement**) – The Magnatech company originated the term as a description of the process utilized in replacing production dialogue in post-production. The proprietary term stuck as a genericized descriptive, which is still in use today. See also **looping.**

DOI: 10.4324/9781003373094-7

AIFF – An acronym for Audio Interchange File Format, developed by Apple Corporation. Native to MacIntosh computers, it's compatible with most modern-day digital audio workstations.

Ambience – Background sounds in sound design; these sounds comprise the atmosphere, or the ambient sounds that clue us into the setting of a scene or location. Also referred to as **atmos**, which is short for "atmospheric" sound – not to be confused with the Dolby surround format known as **ATMOS**.

Ambisonics – Virtual reality sound, in three dimensions, or a synthesized version of three dimensions. Just as virtual reality picture material provides an immersive visual experience, ambisonic sound furthers the illusion by providing the auditory cues necessary to feel like we're present in a three-dimensional world.

Amplitude – A measurement of a sound wave's loudness, as measured in decibels.

Atmos – See **Ambience**.

ATMOS (as in **Dolby ATMOS**) – Not to be confused with Atmos elements in sound design, the Dolby process is a flexible, next-generation surround sound series of formats.

ATSC – The **Advanced Television Systems Committee**. From their website: "[We are] an international body whose mission is to create and foster implementation of voluntary Standards and Recommended Practices to advance terrestrial digital television broadcasting, and to facilitate interoperability with other media." The ATSC was also instrumental in providing technical guidance for audio level standards implemented in the **CALM Act** in the United States in 2010.

Bidirectional – A microphone polar pattern that picks up in two directions on an axis 180° opposite each other. Ideal for two subjects facing each other, but also useful as part of a stereo microphone configuration when used in conjunction with a unidirectional microphone – see also **Cardioid** and **MS Pair**.

Binaural – Denotes the human capability to receive sound via two ears and, from that, extrapolate positional information from the time differences between the two. Our ear–brain combination does a lot of complicated math to tell us the directionality, distance, and intensity of sound, providing the auditory cues that define for us the physical space we're in at any given moment. When applied to a recording style, we're referring to a two-microphone method of capturing sound as naturally as possible to approximate the human hearing experience.

Binaural hearing – How we hear and decipher information about sound as it arrives at our two ears. See **Binaural** above.

Boom microphone – A broad term that could apply to any microphone at the end of a boom pole or mechanical arm. Many microphones might be used in this manner, each with a somewhat specific application. See also **Shotgun**.

Boom operator (sound crew) – Member of the production sound crew who operates the microphone at the end of a boom pole.

Boom pole (also sometimes **fishpole**) – The pole the **boom operator** uses in their work. Depending on the application, they can be shorter or longer, ranging anywhere from a foot or two in length to 20 or 30 feet or more. The best poles will be made of fiberglass or carbon fiber to offset the weight and counterbalance effect that can make them awkward and heavy over longer distances – and longer takes.

BWAV – Adding a "B" to the "WAV" file format, wherein the "B" stands for Broadcast. BWAV files contain all the information of a standard WAV file, with the addition of a timecode address stamp for synchronization purposes later on. WAV and BWAV use the .wav file format, but BWAVs can also be found with the .bwf file indicator. See also **WAV**.

Cable person – See **Third**, **Sound utility**.

CALM Act – In the Commercial Advertisement Loudness Mitigation Act of 2012, the United States Congress directed the Federal Communications Commission (FCC) to establish and enforce rules that initially required commercials to have the same average loudness as the programs they accompany. The loudness standard is now almost universally applied to all television programming and streaming content as a means of quality control, ensuring that all programming is at a consistent loudness level.

Cardioid – One of three general microphone pickup patterns, the cardioid pattern applies to all iterations of the unidirectional pattern – that is, most sensitive on one side of the microphone while rejecting unwanted sound or noise from the less-sensitive side. See also **Omnidirectional** and **Bidirectional**.

Cinema Audio Society (or **CAS**) – Formed in 1964, the Cinema Audio Society is a philanthropic, non-profit organization that has become the organization for sound mixing professionals in the motion picture and television industry. Their goals include fostering community among mixers, educating and informing the general public and the motion picture and television industry about good and effective sound usage, and to achieve deserved recognition for sound mixers as major contributors to the field of entertainment.

Commentative sound – Sound that would otherwise have no business in a particular scene; see example cited in Chapter 3. Commentative sound serves to underline an emotional pretext or to extend the visceral impact within a scene. Also sometimes known as an *internal moment*.

Compander – In signal processing, a combination of two signal processors – a **compressor**, a device to control dynamic range by compressing the peaks of a sound signal, and an **expander**, which works in an opposite manner, suppressing the valleys between sounds. Together, the two can function as a rudimentary noise processor, minimizing background noise while controlling the peaks.

Composer – In visual media, the individual who writes original music for a film or video score.

Compression – In a sound context, this term refers primarily to *dynamic compression*, which is a processor or set of processors that can constrain the dynamic range of an audio signal or program. A mild to moderate compressor will slightly attenuate or soften some audio peaks, making that element more manageable, preventing peak overloads, and helping clarify the sound or intelligibility of the compressed dialogue. A stronger, more constrictive version of dynamic compression is **limiting**.

Compressor – A device that will control the peaks of a sound signal, reducing the overall dynamic range of that signal.

Contextual sound – Sounds that coincide with an on-screen, in-frame action. Dialogue is certainly contextual, but so would be other in-frame actions, such as a door slam, a glass breaking, footsteps, or a dog barking – hence, the old sound cliché "see a dog, hear a dog."

Cue sheet – This can mean different things to different departments, but generally is a means of outlining the work to be done for any given craft. Dialogue and ADR editors might provide a cue sheet of lines of dialogue to be replaced, whereas the Foley crew will create a detailed cue sheet of their tasks. The music team need to create a cue sheet of all music cues – necessary not just for placement, but also for licensing and reporting purposes.

DAW – See **Digital Audio Workstation**.

dB, decibel – A measure of **amplitude** or loudness of a sound. The decibel, more commonly known by its abbreviation dB, is a unit of measurement that is equal to one-tenth of a Bel (B). It can also be used to express the ratio between two values of power. Two signals that differ from each other by one decibel have a power ratio of $10^{1/10}$.

In our world, a decibel measurement either expresses a relative change or an absolute value. Further complicating our relationship with amplitude measurement is the human hearing response, which is far from objective – humans don't perceive relative differences in sound pressure levels in the same way as they are measured.

dB FS – See also **LKFS, LUFS**. In this case, the "full scale" usually refers to a digital meter, which runs from –60 dB (sometimes lower) to 0 dB at the top.

DCP (or **Digital Cinema Package**) – The collection of cinema image, audio, and data streams necessary for theatrical release.

De-click – A part of most **de-noising** plug-in suites, these algorithms can effectively reduce or eliminate transient clicks that might show up in recorded sound. These noises can be organic in nature, such as lip smacks, pops, or other sharp sounds, as well as artificial or other technically based noises, such as digital errors.

Delay – 1. Post-production speak for echo, a plug-in or processor will be utilized to introduce an echo-like delay as part of a sound design treatment. 2. The act of introducing any amount of delay into a track, output, or any sound chain; an example would be putting distant speakers in an auditorium on a delayed signal path to align that sound with the audio coming from the main speakers at the stage or front of the space. See also **Echo**.

Deliverables – The audio elements mandated by the distributor or broadcast outlet taking possession of the final visual story, usually mandated by a schedule of requirements that will outline level specifications, format, and the stems needed to set up the distribution path for both domestic and foreign release.

De-noise – Shorthand for any number of noise correction and audio repair plug-ins on the market. There are several product lines, with Izotope, Waves, and CEDAR being just a few of the industry leaders.

Descriptive sound – Sound from within the world of the scene that serves to underline the location by giving us the appropriate aural information needed. Examples include distant traffic sounds, birds, room tone, and other sounds that originate outside our field of view. See also **Diegetic sound**.

Dialogue normalization (dialnorm) – The practice of getting dialogue to fall into a realistic range while making the other elements – sound effects and music – work in relation to that.

Diegetic sound – Sound that originates from the world of the film, typically dialogue and a majority of the sound design. See also **Non-diegetic sound**.

Digital Audio Workstation (or **DAW**) – A bit of a holdover term, referring to a time when audio workstations were more self-contained, often with proprietary software and hardware incorporated into one system. The term is still used, but has expanded to include any computer-based audio editing and/or mixing program or system used to edit, mix, or otherwise manipulate audio digitally.

Dipped (see also **Undipped**) – The narration has been removed in a dipped soundtrack, but the fades that accommodate it in the music and sound effects remain in place. This can prove problematic with some translations, as different language translations often require different timings, and so a dipped track might create more work for the mixer on the translation end. Consequently, many distributors will require an undipped track, so the sound elements can be tailored to work around the translated dialogue or narration.

DME (also sometimes **MDE, MED**) – The mix, separated out into its stems, or sound family elements, consisting of Dialogue, Music and Effects. When played back together, these should be an accurate representation of the original mix, but the objective is to add the flexibility of removing the dialogue mix so that local language can be dubbed in its place.

Dolby – By itself, this just refers to the company, but Dolby has virtually invented, and certainly refined, many audio processes, including various versions of immersive, or surround sound.

Dubbing – An all-encompassing term that can mean anything from the final mixdown of a piece to the individual mixes of elements within that whole. For example, we might also dub our sound effects down into a composite sound effects stem, streamlining the process toward the final mix. See **Predub**.

Dynamics (or **dynamic range**) – Referring to the range between the quietest and loudest elements of a system's capabilities, or sometimes the variation within a program's mix.

Earwig – A genericized term that most often refers to devices made by the Phonak Corporation as part of their Invisity line of devices. These are wireless, in-ear monitor receivers designed to feed audio to an actor or singer, typically for the purpose of providing audio cues for on-screen action, such as a music track for singing or dancing, or to work in an IFB function for an actor to receive line prompts or action cues. See also **IFB**.

Echo – A discrete repetition of the initial sound, as would happen in an unenclosed space such as a canyon. Shout the word "Echo!" into the void, and one would hear a version of that word come back after reflecting off the distant canyon wall.

Equalization (or **EQ**) – A hardware device or software plug-in designed to affect and manipulate the frequency content of an audio signal.

Expander – An amplitude processor designed to control and suppress the quieter parts of the program's dynamic range, used most effectively to make background noises quieter, which in turn makes the overall dynamic range of a program seem wider.

Fader – Can be a round knob or a sliding fader, sometimes a controlled audio circuit that audio passes through, or (more often these days) a controller of a digital interface that takes those fader moves and converts them into the data necessary to effect changes in audio loudness.

Fishpole – An alternative name for the **boom pole** that the **boom operator** uses in their work.

Fold-down (also **folddown**, or **fold down**) – Taking a surround or other multichannel mix and combining its channels to create a different-format mix with fewer channels, usually two-channel stereo, but sometimes mono.

Foley – See also **Foley artist**. Sound effects, usually the incidental sounds in a scene, that are performed live to picture. The process is named for Jack Foley of Universal Studios, who began adding incidental sound effects that had been so efficiently removed during the production process. "Foley" then became an all-encompassing term to cover all sound effects that were most easily and effectively performed rather than pulled from a sound effects library. The sounds a Foley artist would address would certainly include footsteps – in fact, "Foley" as a term has become almost synonymous with the performance of footsteps to many, with Foley artists sometimes referred to as "**walkers**." A Foley artist is much more than that, however; they will also recreate any sounds caused by objects being handled by the actors on screen, as well as clothing rustle, hand gestures, and other sounds that originate from the body movements of those actors. Foley is necessary to recreate these sounds for the final sound effects **deliverable**, also called the **Fully Filled Effects (FFE)** track, which is an essential requirement if the final piece is to be internationally distributed.

Foley artist – The person who performs any **Foley** work through a variety of means and sound props. Sometimes also referred to as a "**walker**."

Foley mixer – An integral part of the Foley team, the Foley mixer is often much more than just a person at a mixing board. They are tasked with the accurate recording of the Foley artist's work, sometimes combining multiple takes and tracks to create a greater composite sound. They will often assist in editing Foley for the sound editing crew that will be receiving these tracks, helping to set them up with select takes, sync adjustments, and so forth.

Foley shot (or **Foley shoot**) – A Foley team's shorthand for what the rest of us would call a recording session, the process of lining up and recording Foley cues.

Foley stage – Different than other sound, mixing, and recording stages, a Foley stage is purpose-built for the work at hand. Often built with "pits," or openings in the floor for soft surfaces like sand, gravel, rock, and so forth, but with removable surfaces for anything else the Foley artist might need to match on screen, including different types of wood, tile, steel, and nearly anything else one might imagine as a walkable surface. There will also be shelving and storage for the wide selection of Foley props the Foley artist might need, and some stages might have plumbing and different-sized water tanks for wet Foley effects.

Follow cart – A production sound crew's utility cart that will carry everything from spare microphones and cables to batteries and boom poles, in addition to any number of the items needed to help do the work of production sound. The follow cart will usually be stationed near the set, but in a more out-of-the-way location, serving as the supply unit for the sound crew's needs.

Frequency – In sound, a measure of the number of times per second a full sound wave cycle occurs, measured in cycles per second, or Hertz. Humans can generally hear many of the frequencies from 30–40 Hertz to 17,000 Hertz (or 17 kilohertz), although some individuals might be able to perceive even higher frequencies. For this reason, we often round down to 20 Hertz on the low end of the scale and up to 20,000 Hertz – knowing that those two measures generally go beyond the range of most humans. See also **Frequency response**.

Frequency masking – When the frequencies present in unwanted sound – or noise – interfere with similar frequency ranges in the sound we want. Broad frequency range noises, such as air handling equipment, wind, water, traffic, and so forth, are the most difficult to deal with.

Frequency response – A measurement of the capabilities of a system for reproduction of the frequency spectrum. A "system" can include any gear component or piece of equipment that is designed to pass, generate, or reproduce sound, but can also include human hearing response.

Fully Filled Effects (or **FFE**) – Part of the deliverable requirement, usually for international distribution, where the dialogue track will be replaced with local language. When that production track is not being used, all the incidental sounds that were a part of it will no longer be present, requiring them to be re-performed by the Foley team. Yes, it is often like doing the film twice.

Fundamental Tone (or **Fundamental**) – The strongest frequency in a multi-frequency sound, often the bass or lowest-frequency element present.

h.264 – Also known as MPEG-4 Part 10, this is a compressed video format typically used as the video program reference, or "work video" in audio post-production platforms, due to its smaller file size.

Hard effects – Sound effects that are sync-critical, such as a door slam – we can see the point of contact, where the door hits the door jamb, so its sound should coincide and synchronize with that. See also **Soft effects**.

IATSE (or the **International Alliance of Theatrical Stage Employees**) – Represents freelance and staff workers in the stage, motion picture, and television production and post-production industry throughout the United States and Canada. There are more than 500 affiliated local and nationwide union chapters and guilds that comprise IATSE, a national labor organization with 104,000-plus members.

IFB (or **Interruptible Fold Back system**) – Used to communicate with talent or members of a production team via closed, one-way radio. Often used in television production studios, where a producer may need to talk to the on-air talent directly, it is also utilized on some film sets so that the members of the sound department can hear and respond to the mixer's direction.

Lavaliere – See **Wireless**.

LFE (or **Low Frequency Emitter**) – Refers to the subwoofer channel in a surround mix; the ".1" in 5.1, it is the speaker that outputs the channel with the low bass element of the mix. See **subwoofer**.

Library – Referring to licensable music or sound effects for use in a production's soundtrack. Music libraries typically require a licensing process before the use of any cuts is allowed; sound effects libraries, once purchased, generally provide unlimited use of those effects.

Limiter – A heavier form of **compression**, where an audio signal is constrained to the threshold ceiling set on the processor plug-in. Where compression takes a softer, more subtle approach, a limiter's threshold becomes a barrier – no dynamic peaks can surpass the limiter's threshold. This has its applications in broadcast but is useful as a kind of backstop for possible peaks that might surpass the rest of the program.

LKFS (also **LUFS**) – It is a not-quite acronym for **Loudness, K-weighted, relative to Full Scale**. It is foremost an internationally agreed-upon scale of measurement. There is a movement afoot – primarily from the European Broadcast Union (EBU) – that we should instead opt for the more concise **LUFS**

(Loudness Units, Full Scale) as the more intuitive term. **LKFS** was adopted by the International Telecommunication Union (ITU) as the standard in the level measurement specification ITU-R BS.1770-3, the most recent iteration of loudness measurement for programming and the like. It should be noted that both **LKFS** and **LUFS** mean exactly the same thing; the only difference is the moniker. In each case, one *Loudness Unit* equals one decibel.

Looping – See also **ADR**. The process of replacing production dialogue in post-production. The term "looping" is a holdover from the film days, when editors would cut a loop of film for the scene they were working on – the loop would play over and over as the actor did and redid the line, with all takes being recorded.

LtRt – A folded-down stereo version derived from the previous surround format the film was originally mixed in. Stereo tracks are provided alongside the surround for compatibility with different markets, outlets, broadcasters, streamers, and viewers who don't have surround listening capability.

LUFS – See **LKFS**.

Martini – On a film set, meaning the last shot of the day.

Master fader (or **master**) – In a mixing or recording console, the final output of that console's mixdown. That mixdown can serve any number of purposes – it could be the final mix of a program, or a monitor feed for a production mixer to listen to all tracks at once. There are also variations on the **master fader** theme, see also **Submaster**.

Mickey-mousing – Typically a derogatory term for a style of sound effects editing, meaning that a sound editor has gone a little overboard with their sound effects treatment, there are too many sound effects in place, and they're proving to be distracting.

Mix stage – The place where the magic happens and all the individual tracks come together to help create a final **mixdown** of all the sound editing that has been done, along with the edited dialogue and final music score.

Mixdown (also sometimes just **mix**) – The result of all that hard post-production sound work, where all the sound elements are combined into a single, coherent track for the final version of the visual story. That mixdown track can be any one of several surround formats, 2-channel **stereo**, or 1-channel **mono**.

Mixer – This one depends on context. 1. During production on set, the mixer is the Sound Department's head and the one tasked with getting clean audio (most often, just dialogue). 2. The mixer position also exists in the post-production world, where this can be one of a team of mixers mixing the piece, an individual doing the same, or any number of specialists along the way, e.g., the Foley mixer, ADR mixer, etc. 3. The term can also be applied to hardware, meaning the mixing console or work surface.

Mixing – The process of combining two or more sound elements together to make a coherent composite of the original elements. This can be on the largest scale, such as a feature film mix, or within that process where we simply combine like elements together to help streamline the final mix – for example, combining multiple sound elements to create a larger-than-life composite sound, such as a car crash. There would be elements of metal, glass, impacts, tires screeching, and so forth to create a dramatic whole sound – but we can help the final mix along by "locking in" the relationships between those individual sounds, thus taking some of the mixing decisions off the table for the always-complicated final mix.

Monitor(s) – Often taken to mean the screen we see the visual element or computer output on, in the sound world it can also apply to the physical speakers we're listening to (monitors), or it can also be a verb, as in "to monitor" our various audio outputs.

Mono (or **monoaural**) – Applied to mixdown, meaning the audio output here, or mix, has only one combined channel.

MOS – Short for "Mit Out Sound," which dates to the old days of Hollywood. The story goes that a German director told his camera assistant that they would be rolling without sound – or, as the director said, "mit out sound." The assistant dutifully wrote "MOS" on the slate, and the term has stuck. MOS shots are universally understood to have been shot without sound rolling.

MP3 – Technically, the term is shorthand for **MPEG-1 Audio Layer III** or **MPEG-2 Audio Layer III**. This is a compression format for encoding audio that, like other forms of data compression, uses approximation and interpolation to create a representative, yet smaller, version of the original audio file. Considered insufficient for professional sound use, the format endures as a delivery medium because of its smaller size and relatively good fidelity.

MP4 (or **MPEG-4 Part 14**) – **MP4** is a digital media container format most commonly used to store video (and/or audio). It is a convenient video format for use in audio post-production because of its relatively small size, and like most other container formats, it allows for streaming online.

MPEG – Confusingly, there are two bodies that go by this acronym. 1. The **Moving Picture Experts Group** (**MPEG**) is an alliance of groups that sets standards for media coding, including compression and encoding of audio, video, and graphics. 2. The **Motion Picture Editors Guild** (**MPEG**, **IATSE**

Local 700) is a superlocal affiliate of the International Alliance of Theatrical Stage Employees, representing freelance and staff motion picture film, television, and sound editors – among other post-production professionals and story analysts throughout the United States and Canada. The Motion Picture Editors Guild (Union Local 700) is one of the 500 affiliated local unions that comprise IATSE, a national labor organization with 104,000-plus members. There are more than 6,000 members of MPEG 700.

MS Pair – Short for Mid-Side Pair, a stereo microphone technique that combines the patterns of two microphones – one with a cardioid pattern, the other with a bidirectional pattern. When properly combined, the resultant stereo recording is phase-accurate and adjustable after the fact.

Multiband – Any instance where multiple ranges of frequencies (bands) are being affected or processed, such as in a multiband compressor or equalizer. In each instance, the frequency bands are divided into ranges or regions that can then be processed independently of each other.

Multitrack – Meaning only more than one audio track in a file or on tape. Many tracks can be synchronously recorded and stored so that their relationship to each other is essentially locked into place.

Music Department – On a film, the Music Department can be wide-ranging, to include the **Music supervisor, Music editor, Composer**, and all other personnel who help see the music track, or score, to the final mix.

Music editor – The person tasked with placing and (if necessary) editing the music tracks to fit the scene for which they were designed, in the case of original music score, or editing other existing music cues to work with the scene. In all cases, the music editor is charged with preparing all music elements for the final mix stage.

Music library (also **stock music**) – Many music libraries are available for hire, where one can find a piece of music that will work with a particular visual story. Seen by many as a compromise over the more bespoke, original score a composer might provide, music libraries provide an effective and often more affordable option.

Music supervisor – The person charged with curating prerecorded music to be placed into a film's soundtrack. Depending on the style of music and the film itself, the process often requires a music supervisor with an almost encyclopedic knowledge of music and styles.

Noise gate (or sometimes just **gate**) – The opposite of a dynamic **compressor** or **limiter**, a noise gate is a close relative to the dynamic **expander**, pushing the quiet parts of a sound signal to be even quieter. When used well, it can help to reduce distracting background noise by pushing the noise floor to lower, quieter levels.

Noise processing (or **noise reduction**, see also **De-noise**) – There are many modes of noise suppression, ranging from single-ended to double-ended encoded systems (such as analog dbx and some Dolby systems) to the intelligent plug-ins used today. Today's noise processors are much more powerful because they can remedy and mitigate a wide variety of noise issues, from broadband noise (such as traffic, rain, or air handling noise) to more invasive sounds like squeaks and rattles. They can often minimize unwanted reverberations and short-duration, transient sounds like clicks.

Non-diegetic sound – Parts of the soundtrack that come from outside the world of the film, see also **Diegetic sound**. Non-diegetic sound is most often narration and musical score, although it can also have elements of sound design in it. Most typically, it is sound that isn't perceived or heard by the characters in their film scene world.

Omnidirectional – Referring to a microphone's pickup pattern, an omnidirectional microphone picks up sound equally in 360°. See also **Cardioid** and **Bidirectional**.

Pickup (or **pickup pattern**) – Refers to the polarity pattern of any given microphone. Microphones generally fall into one of three general pickup pattern types: omnidirectional, picking up sound in 360°; unidirectional, which has a polar pattern with greatest sensitivity in front of the mic; and bidirectional, which picks up equally in two directions 180° opposite each other (see Figure 1.23).

Playback – Referring to any mode of playing back audio or video but can mean differing things in different contexts. On set, we have video playback, or video assist, who are there to provide playback of takes for the director. We might also have sound playback, who would provide a reference track – usually music – for the actors to act or react to, such as singing or dancing to a prerecorded song.

PolyWAV – A multichannel WAV audio file format, wherein multiple audio tracks can be interleaved into one large file instead of many mono files. See also **WAV**.

Predub – The process of premixing related elements in preparation for the final mix to make that process more streamlined and a little less complicated. For example, a dialogue predub might consist of making some preliminary mix decisions around processing and effects on elements of the dialogue track, like noise processing, reverb treatments, and certainly equalization and compression. These elements will be provided as stems that are shipped to the mixing stage for the final mixdown.

Pre-lay – An early stage of sound design, where sound effects – often placeholders – are edited to picture as a sort of sketch of ideas to come.

Preproduction – The planning process that precedes the film production, wherein all involved parties get together to work out the technical and logistical challenges of producing the scenes to come.

Previz – A pre-production tool, usually animated, to provide proof of concept and help plan the production to come. Previz pieces can be elaborately animated or very rough but often require some sound design to help make the point. See also **Storyboarding**.

Pro Tools – One of many Digital Audio Workstations (or DAWs) being used in the field of audio post-production. Protools is an AVID product and likely has the greatest market share among post-production professionals, but there are many other DAWs in use, such as Adobe's Audition product, Nuendo, Logic, and more.

ProRes – An Apple video format, high-quality and "visually lossless" lossy video, capable of up to 8K resolution.

Proximity effect – Most pronounced in unidirectional, dynamic microphones; this is a phenomenon where the bass response on the microphone is exaggerated when the source is close to the mic. This can be both useful and detrimental, depending on the situation – on-air personalities will use the proximity effect to make their voices deeper and gain bass. But in some situations, it can also sound unnatural, such as when recording dialogue for a scene – pronounced bass simply doesn't match perspective.

QuickTime – Apple's video file format, usually compressed. See also **MP4**.

Radio Frequency Interference (or **RFI**, or sometimes just **RF**) – Unwanted noise interference induced over cables and their connections, picked up and intermingled with the clean audio signal. Difficult to get rid of in post, the best course is prevention and mitigation at the source. Inferior or distressed cabling is often a cause, but there are other mitigations available, such as cable chokes, transformers, and electronic isolation devices to help with the issue when it arises.

Reverberation (or **reverb**) – Both an acoustic phenomenon and an electronic signal processing mode. Acoustically, it can be an unwanted element when on set, affecting our ability to record clean, clear dialogue. In post-production, reverberation can help establish the space around dialogues or sound effects by making them sound like they are in the environment we see in the scene.

RFI – See **Radio Frequency Interference**.

Scout (also **location scout**) – A necessary process ahead of any shoot to establish logistics and the look of the upcoming shoot. Sound should be present during the secondary scout so they can be aware of any noise issues ahead of time, such as being near a busy road or highway or heavy air handling noise. We might not be able to mitigate all noises but can at least adjust expectations before concerns are raised.

SFX – See **Sound effects**. Not to be confused with the visual term *special effects*.

Shotgun – Shorthand for a very directional boom microphone used on set. In the extreme, it is a unidirectional microphone with a far greater side rejection and, consequently, "reach" than the standard unidirectional pattern, making it ideal for the distances and conditions experienced on a set.

Soft effects – Sound effects that don't have a defined sync point, or maybe that have an offset sync point that isn't as critical to its placement next to visual elements. One example would be a car passing by our perspective – the point of sync is somewhat less-defined and comes sometime after the onset of the sound itself. The only sync is the point where that car passes us, but there could well be a few frames of "play" in there – it is certainly not as sync-critical as a door slam, which would be an example of a **hard effect**.

Sound design – The reason we're here and the reason you're reading this text! Sound design is the overall approach to the soundtrack to help it add to the overall story, taking a planned, aesthetic approach to applying sounds to the finished result. Every sound placed has an opportunity to help tell the story.

Sound effects (also **SFX**) – Any non-dialogue, non-music sound used in the design and execution of the soundtrack. It can encompass sounds from various sources – field/production audio, Foley, sound effects libraries, or the sound editor's proprietary library of original effects.

Sound effects library (also **stock effects**) – There are a number of purchasable sound effects libraries, as well as some crowd-sourced databases. Many of the higher-end (read: more expensive) libraries are professionally recorded and edited, ready to plug and play.

Sound Pressure Level (SPL) – A measurement of loudness (**decibels**) in the acoustic domain – as opposed to other decibel measurements in the more theoretical domain of technical response of the gear we use, such as microphones, speakers, and so forth. SPL is how we measure sound level in the environment.

Sound utility (sound crew) – Also known as the **cable person** or **third**. This person is typically the third member of the production sound crew, tasked with the minutiae of keeping the Sound Department on track, in communication with the other departments, and fully functional. The "cable" reference is from the number of cables and wires this position once had to manage, most often by keeping a position behind the **boom operator** to keep them free of tangles and making those cables less of a safety hazard. This position is also tasked with placing wireless mics on the talent, keeping the equipment charged and batteries fresh in all the (now mostly wireless) microphone transmitters and receivers, organizing Sound Department logs and other paperwork, moving the **sound cart, follow cart**, and other sound-related gear into position, and working as a second boom op when the situation requires it.

Soundcart (also **sound cart**) – The production mixer's cart that contains all their equipment for recording on-set audio. Typically, it consists of a recording device, a mixer, video monitors, antennae and multiple audio and video connections, and a power distribution system with a battery backup. A mixer's soundcart is tailored to their individual needs, style, and method of working; they are not the same. See also **Follow cart**.

Soundtrack – The overall sound treatment for a video or film. Many people equate the term with just the music element of a film, but the term – for sound designers and mixers, anyway – is all-encompassing and includes all the sound work and elements necessary for a fully realized sound story.

Spotting – See also **Cue sheet**. The process of mapping out the work to be done in post-production, be it laying out Foley work, dialogue to be replaced in ADR, or in a broader context with the director and Music Department, where the sound supervisor will participate to map out the coverage of sound design vs. score.

Stems – A **mixdown** of related elements into a composite track. Generally, the mixing process is a series of adjustments and smaller mixes, where relative levels are established in smaller, easier-to-manage chunks.

Stereo – A throwback to the stereophonic heyday, typically taken to mean two-channel, left and right output, but technically meaning more than one channel of audio. For instance, Dolby Stereo is a surround format that includes left, right, and center front channels and a mono surround channel, or LRCS.

Storyboarding – A pre-production tool to help plan shots and subsequent edits. Usually consisting of static drawings and images, as compared to **previz** presentations, which are often animated. Storyboards can be helpful for the upcoming production and planning of a shot list.

Streamers – Often used as an all-inclusive term for all the online streaming channels (Netflix, Hulu, Prime, etc.), it also refers to a visual reference system for cueing purposes in audio post-production. One example would be for ADR sessions, where a visual series of lines of numbers is used to cue the actor to redo their line in the right place. Composers will also use a system of visual streamers to cue their conductor and/or orchestra to start playing at the right point for a music cue.

Subgroup – Usually a function to help facilitate the mixing process, where like sounds are combined into groups. These can be as wide-ranging as an all-inclusive sound effects group, to a group of just Foley, and so forth.

Submaster – A mix is usually fed to its output through one point, controlled by a master fader. A submaster is a grouping, usually of the "sound macros" of dialogue, music, and effects. This gives us separate control over each group but also allows separate output for each deliverable. See also **Deliverables, Master fader**.

Submix – A combination of similar elements grouped and mixed as a smaller subset of the larger track build. One example might be combining all sound effects tracks together, balancing them out against each other in preparation for the final mix. Submixes can be more granular, however, and we might get into a submix of sound effects to create a larger, composite sound, like an explosion – there could be several layers there, all in the name of making a bigger, more impactful final sound effect.

Subwoofer – A larger speaker designed to play only low bass frequencies, typically below 80 Hz. It has applications in post-production and distribution, where it is part of the sound playback system (the .1 in 5.1). A subwoofer can also be used on set to provide a "**thumper** track" for actors to sync to when singing or dancing. The low-frequency nature of the thumper allows the mixer to simply filter out those sub-bass frequencies and preserve the integrity of a clean dialogue track.

Surround (or **Surround Sound**) – There are many surround formats and options for immersive sounds these days, so it's an inclusive term that requires context and further definition to be used properly. As an example, most DVD and some broadcast programs are mixed in 5.1, but there are many other formats and applications, including 7.1, SDDS, Dolby ATMOS – the list goes on.

Temporal masking – When sound, in the form of unwanted noise, is louder than and obscures the sound we are trying to hear or record. An example would be a door slamming over a line of dialogue on set, where the louder door sound would likely interfere with our ability to fully hear and understand the word of dialogue where it happened.

Temporal shift – The loss of hearing sensitivity that occurs in our hearing over time. As our day progresses, we become less sensitive to loudness and may compensate by increasing the loudness of the things we listen to. A common example would be how our earbuds or car radio can seem much louder in the morning when we first turn them on than we remember them being the night before. This happened because we became less sensitive to loudness over the course of the previous day, but listening with rested ears in the morning showed us how loud things really were the night before.

Third (sound crew) – See **Sound utility**, **Cable person**.

Thumper – See **Subwoofer**.

THX – A specification designed to ensure the moviegoing public would experience a film in the same manner as it appeared before the director on the mixing stage. Every aspect of the theater is examined and specified so that the audience hears the film as the mix team intended.

TPK (or **True Peak**) – A measurement of loudness of any peaks in the program material, including inter-sample peaks – hence, the term "true." Most distribution outlets have strict requirements in this area, as material that surpasses an acceptable true peak will likely cause distortion and possibly a loss of data.

Track – Referring to any number of ways of delivery within an audio workstation system, as part of a video file, or as part of the final sound deliverable.

Transducer – A process by which energy is converted from one form to another – for example, the transducer in a microphone converts acoustic energy into an electrical signal that can be amplified, recorded, or manipulated in any number of different ways. Conversely, the transducer in a speaker converts that electrical signal back into acoustic energy.

Transient – An unwanted click or pop that can appear in a sound signal. They can sometimes be organic, such as lip and mouth sounds, props or other objects on set, or artifacts or errors in the recording process. **De-noising** software – specifically, **de-clicking** algorithms – can effectively eliminate or reduce these artifacts when they show up.

Tree – Production sound shorthand for the antenna array where all the microphone receivers are mounted. The tree will be placed as close to set as possible to minimize the potential for RFI (**radio frequency interference**, or sometimes just **RF**) getting into the microphone's frequencies. It is then physically wired to the soundcart.

Undipped (see also **Dipped**) – A deliverable requirement often made of documentary mixes, where the mix is smoothed out with the original narration removed. This facilitates easier replacement with local-language narration later on, as different translations have different timings. This way, the track can be delivered without the constraints of the fades in background material that were made to accommodate the original narration track.

Unidirectional (see also **Cardioid**) – Usually referring to the polarity of a microphone, meaning the microphone has an area of greatest sensitivity at 0°, or in front of the microphone, with the greatest rejection of sound at the 180° position opposite. There are many different types and refinements of the unidirectional design, ranging from fairly wide pickup patterns to very narrow patterns, such as those utilized in boom microphones on set, commonly referred to as **shotgun** microphones.

Velocity – We can measure the speed of a sound wave in a medium – usually air – to better understand the acoustical and physical implications. Sound is significantly slower than light, and with enough distance, this phenomenon can be easily noticed – imagine a fireworks display, where the sound of the report is significantly later than the visual of the explosion.

VR – Acronym for Virtual Reality, sometimes to include the audio, which would be more properly called by any number of trade names and processes, such as Ambisonic, or spatial.

Walker – Another name sometimes used to describe **Foley artists**.

WAV (or **Waveform Audio File Format**) – Developed initially by IBM and Microsoft for the purposes of storing uncompressed audio on PC. Now widely adopted by the sound industry, it is a standard file format for storing and manipulating professional audio. It can accommodate a variety of track counts, sample rates, and bit depths, making it a flexible option for professional audio applications.

Wavelength – One part of the measurement of a sound wave, we can measure the physical distance a sound wave travels over the course of one complete cycle of the wave.

Wire – (v), as in "to wire," or place a lavaliere (wired or unwired) on an actor or talent in preparation for a shoot. "I asked my third to *wire* the actors before they come on set."

Wireless – Sound crew shorthand for a wireless connection to the microphones being used on set, usually specifically the lavaliere. Confusingly, lavalieres are sometimes also referred to as "wires," whether wired or wireless.

Wires – (n), shorthand on set for lavalieres and their transmitter/receiver setup.

Workstation – In the context of sound work, it's less about the physical equipment than the software itself, which can be installed on virtually any computer to do the work of audio editing, assembly, and mixing.

XLR – Itself short for External Line Return, this is a balanced connection for professional audio, consisting of a positive (hot) pin, a negative pin, and a third for ground.

6.2. REFERENCES AND OTHER SOURCE MATERIAL

6.2.1. Educational and Informational

www.filmsound.org – A clearinghouse for articles, interviews, and other sound design resources.

www.nofilmschool.com – Seriously, can't get enough of this site. Perfect for the DIY filmmaker.

6.2.2. Sound Guilds and Unions

The Cinema Audio Society, www.cinemaaudiosociety.org. CAS is one of the guilds that has voting rights in the Academy Awards and the Emmys; it's also a great resource for educational material and networking.

International Alliance of Theatrical Stage Employees (IATSE.net) is the union that works in all forms of live theater, motion picture and television production, trade shows and exhibitions, television broadcasting, and concerts as well as the equipment and construction shops that support all these areas of the entertainment industry. More than 150,000 workers strong in virtually all arts, media, and entertainment crafts, their mission is to improve all entertainment workers' lives both inside and outside the workplace. Many of the IATSE locals serve a particular state or region, but there are a couple of exceptions.

IATSE 600, the **International Cinematographer's Guild**, or **ICG**, is the Super Local (meaning it covers all IATSE jurisdictions) serving camera professionals. https://www.icg600.com/

IATSE 695 is the LA-based local that serves production sound people, video assist, and projection personnel. https://www.local695.com/

IATSE 700, Motion Picture Editors Guild (MPEG) – www.editorsguild.com. As an IATSE Super Local, MPEG 700 covers all the visual editing crafts and post-production sound categories for mixing and editing.

The Motion Picture Sound Editors (**MPSE**) is another group that has nomination and voting rights in the Academy Awards and Emmys (among others), as well as its own awards. Through the efforts of its volunteer board of directors and dedicated international membership, it strives to celebrate community, and the creative and technical advances in sound.

6.2.3. Sound Effects Libraries

Hollywood Edge – Sound effects library available for purchase and by subscription, https://www.hollywoodedge.com/

Pro Sound Effects – One of the largest sound effects library collections available for purchase or subscription, https://www.prosoundeffects.com/

Sound Ideas – Another sound effects library for purchase or subscription, https://www.sound-ideas.com/

Index

Note: page numbers in *italic* type refer to figures; those in **bold** type refer to tables.

5.1 format 63, **66**, 105–106, 176, 183
5.1+2 format 183
7.1 format 63, **66**, **67**, 105–106
16-bit audio recording 52
24-bit audio recording 52–53
32-bit audio recording 53–54
"40 Hz room", the 30–31

Ableton 183
acoustics 31; HHR 39–42; perception *38*, 38–39; phase *31*, 31–33, *32*, *33*, *34*; reality and psychology of sound *38*, 38–42; reflections 34–37, *36*
actors, and noise reduction 24, 138
Adobe Audition 109, *159*, 183
Adobe Premiere 109
ADR (Automated Dialogue Replacement) 93, 105, 109, 124, 127, 138–143, *140*, *142*, 183; VO Booth and ADR stage 109–112, *110*; *see also* looping
Advanced Television Systems Committee (ATSC) *see* ATSC (Advanced Television Systems Committee)
advertisements, loudness issues 167–169, 170–171, 185
AI (artificial intelligence) 62, 179, 181–182
AIFF (Audio Interchange File Format) 184
Altman, Robert 93, 139–140
Amazon Prime 63, 175, 178
ambience 184
ambient sounds 98, 99, 112, 116, 119, *121*, 124, 129, 150, 155
ambisonics 162, 184
ambo 99
American Conservatory Theater 5
amplitude 12, *13*, 13–17, 184, 186
amplitude processors 58–60, *59*, *60*
Apocalypse Now 5–6, *6*
Apple 163, 175; AIFF (Audio Interchange File Format) 184; ProRes 196; QuickTime file format 196
artificial intelligence (AI) 62, 179, 181–182
assistant directors 85, 86, 143–144
atmos 99, 138, 183
ATMOS 63, **66**, **67**, 105, 106, 164, 183
ATSC (Advanced Television Systems Committee) 168, 183; A/85:2013 specification 168
Audio Interchange File Format (AIFF) 184
audio playback 74
Automated Dialogue Replacement (ADR) *see* ADR (Automated Dialogue Replacement)
Avantone mix cubes *173*
Avid S1 console *159*
AVID technologies 104

bag rigs 76–77
BaseHead 149
Beckett, Ray 29–30
Bel 13, 186; *see also* dBs (decibels)
Bell, Alexander Graham 13
bidirectional pickup pattern 46, *46*, 47, 184, 185
binaural 184
binaural hearing 35, 39–41, 185
bit rate 52–53, *53*
Blade Runner 123
Blume, Tara *110*, *111*
Bogart, Humphrey 181–182
boom microphones 51, 74, 78, *79*, 79–80, *80*, 85, 185; phase interference 33, 34; *see also* shotgun microphones
boom operator 69, 70, 72–73, 81, 82, 86, 185, 189, 198
boom poles 47, 69, 185; *see also* fishpole
BWAV file format 185; *see also* WAV file format

cable person *see* third (cable person, sound utility)
call sheets 68, 69, 85
CALM (Commercial Advertisement Loudness Mitigation) Act 2012, US 168, 176, 184, 185
camera-mounted microphones, phase interference 33
captioned films 132
cardioid (unidirectional) pickup (polar) pattern *46*, 46–47, 48, 184, 195, 201
CAS (Cinema Audio Society) 185, 202
CBS 178
CDs 162, 163
chorus 62
Cinema Audio Society (CAS) 185, 202
Cinemax 178
"comb filtering" 32–33, *33*
commentative sound 120–121, *121*, 130, 186
Commercial Advertisement Loudness Mitigation (CALM) Act 2012, US 167, 168, 176, 184, 185
compander 186
composers 133, 134, 151, 152–154, 156, 158, 186, 194
compression *10*, 11, *60*, 186, 191
compressors 59, *59*, 186, 194
concert halls; velocity issues 26, *26*; *see also*theater sound
condenser (capacitor) microphones 43–45, *44*, *45*
contextual sound 117–118, *118*, 146, 186
crowdsourced content 175
cue sheets 146, 147, *147*, 148, 186; *see also* spotting

DAWs (digital audio workstations) 54–56, *55*, 114, 182, 188; portable 104, 106; *see also* workstations
dB FS 187; *see also* LKFS; LUFS
dBs (decibels) 13–14, 186–187; dB sound pressure level (SPL) 14
DCP (Digital Cinema Package) 187
de-click 187, 200
delay 187; *see also* echo
delegation 154–155
deliverables 187, 189; and formatting 171–179, *172*, *173*
de-noise 187, 200; *see also* noise processing/reduction
descriptive sound 119–120, *120*, 125, 187
dialogue 70, 87, 116, 117; amplitude processors 58–59; human voice recording 80, *80*; intelligibility 125, 126–129, 161–162; post-production 50–51; VO Booth and ADR stage 109–112, *112*; *see also* ADR (Automated Dialogue Replacement); DME (MDE, MED) (Dialogue, Music and Effects)
dialogue editing 90, 92–93, 123, 126–127, *128*, 129, 137–144, *140*, *141*; delegation 154–155
dialogue normalization (dialnorm) 187
dialogue replacement *see* ADR (Automated Dialogue Replacement)
dialogue tracks 2, 70, 97; track count 105
diegetic sound 116, 187; *see also* descriptive sound; non-diegetic sound
diffuse reflections 36–37
digital audio workstations *see* DAWs (digital audio workstations)
Digital Cinema Package (DCP) 187
dipped 188; *see also* undipped
direct sound 35, 36, *36*

directors 28, 50, 68, 69, 70, *72*, *75*, 86, 91, 92, 122, 123, 127, 152–153, 154, 160; and dialogue replacement 138, 139; and music 133–135; and sound supervisors 153, 154
discrete/independent reflections 36
distribution; distribution formats 174–175; evaluation and specifications 166–171, *169*; formatting and deliverables 171–179, *172*, *173*; mixing and mastering for the medium 161–171, *169*, 180; systems and media 62–64, **64–67**
dither 52
DME (MDE, MED) (Dialogue, Music and Effects) 105, 106, 135, 176–177, 188
Dolby **64**, 168, 188
DPA 4660 microphone 45, *46*, *79*
dubbing 96, 97, 103, 188; dubbing/mixing stages 112–114, *113*
Dugan, Dan 5
Dune 89–90, 123–124, 152, 165
dynamics (dynamic range) 14, 188

earwigs 74, 188
echo 36, 37, 61, 62, 188; *see also* delay
editorial team 89, 90
Electric Department 70
electronic sound chain, frequency range 12
environmental conditions, and velocity 25
environmental noise 15–16, 22, 62, 124–125; selective hearing 40–41
envisioning sound 153–160
EQ (equalization) 189
EQs (equalizers) 57–58, *58*
equipment 5, 42–62, 74–85, *75*, *76*, *77*, *79*, *80*, *82*, *83*, 104–114, *107*, *108*, *110*, *111*, *112*, *113*, 120; VO Booth and ADR stage 109–112, *112*; *see also* DAWs (digital audio workstations); microphones; production recorders; signal processors and plug-ins
Ethernet 75
expanders 60, 186, 189, 194
External Line Return (XLR) 202

faders 56, 59, 100–101, *101*, 189
"fake news" 182
FCC (Federal Communications Commission) 167, 185
FFE (Fully Filled Effects) 97, 132–133, 189, 190
field recorders *see* production recorders
file formats 104–105; formatting and deliverables 171–179, *172*, *173*
films; sound distribution needs 177–178; velocity issues 27, *29*, 29–30
fishpole 185, 189; *see also* boom pole
flanging effects 62
FM radio transmitters 173
fold-down (folddown, fold down) 189
Foley 94–97, *95*, 98, 105, 109, 113, 129, 130, 131, 145–147, *145*, *147*, 148, 149, 189; contextual sound 115, 116, *118*; Foley artists 90, 126, 146, 147, 154, 189; Foley mixer 189; Foley shot/shoot 90, 109, 190; Foley stage *95*, 95–96, 109, *110*, *111*, 145, *145*, 146; gun-handling sounds 96
Foley, Jack 94
follow carts 82–83, *83*, 190, 198; *see also* sound carts
footsteps 95, 130, 145, 147, 189
formatting; and deliverables 171–179, *172*, *173*; *see also* file formats
Formosa Hollywood Stage 1 156, *157*
"fourth wall" 4, 116
frequency *10*, 10–12, *11*, 190; and wavelength 30–31
frequency response 17–22, *18*, *19*, *20*, *21*, 190; masking 22–24, 190
Fully Filled Effects (FFE) 97, 132–133, 189, 190
fundamental/fundamental tone 191
"futzed" tracks 137

gaming 179
gate *see* noise gate
gear *see* equipment
"golden ear" 163
Grenga, Ed 134–135
gun-handling sounds 96

h.264 (MPEG-4 Part 10) 191
hard effects 98, 129, 149, 155, 191, 197
harmonics 11–12
HBO 178

headphones 71, 175; noise-cancelling 31–32, *32*
hearing *see* HHR (human hearing response)
Hertz 12
HHR (human hearing response) 9; amplitude range 14, *38*, 38–39; binaural and selective hearing 35, 39–41; frequency response 12, 16, 17–22, *18*, *19*, *20*, *21*, 38, 41; hearing loss 14–17, **16**; strengths and weaknesses 41–42
Hollywood Edge 149
horse sounds 129, 149
Hulu 63, 178
human hearing *see* HHR (human hearing response)
human voice 127; non-vocal human sounds 138; recording of 80, *80*; *see also* dialogue
Hurt Locker, The *29*, 29–30

IATSE (International Alliance of Theatrical Stage Employees) 191, 193–194, 202; IATSE 695 202; ICG (International Cinematographer's Guild), IATSE 600 202; MPEG (MotionPicture Editors Guild), IATSE Local 700 202
IBM 201
ICG (International Cinematographer's Guild), IATSE 600 202
IFB (Interruptable Fold Back System) 81–82, 191
immersive sound 174
incidental sounds 90, 143, 145
indirect/reflected sound 35–36, *36*
Instagram 7
intelligibility, of dialogue 127, 129, 161–162
internal moments 186; *see also* commentative sound
International Alliance of Theatrical Stage Employees (IATSE) *see* IATSE (International Alliance of Theatrical Stage Employees)
International Cinematographer's Guild (ICG), IATSE 600 202
international distribution; dialogue translation and replacement 126, 176–177, 188, 190, 201; FFE (Fully Filled Effects) 97, 132–133, 190; Foley effects 145; music 126; sound effects 137–138, 142–143, 176
Interruptable Fold Back System (IFB) 81–82, 191
Interstellar 160, *160*
ITU (International Telecommunication Union), ITU-R BS.1770-3 192
Izotope 138; Insight 2 level metering plug-in *169*

King, Richard 160
King, Stephen 180

language, spoken 39–40; *see also* dialogue
lavaliere microphones 24, 45, *46*, 47, 48, 51, 78–79, *79*, 80–81, 84; phase interference 33, 34
LFE (low frequency emitter) 191
Lifetime 178
light, velocity of 25–30, *26*, *27*, *29*
Lighting Department 70
limiters 59–60, *60*, 191
limiting 186; *see also* compression
lip reading 40–41
LKFS (Loudness, K-weighted, relative to Full Scale) 191–192
location scouts *see* scouts (location scouts)
location scouts (scouts) 197
looping 127, 138, 192; *see also* ADR (Automated Dialogue Replacement)
loudness 186; of advertisements 185; amplitude processors 58–60, *59*, *60*; LKFS (Loudness, K-weighted, relative to Full Scale) 191–192; LUFS (Loudness Units, Full Scale) 191–192; metering *169*, 169–170; SPL (sound pressure level) 197; *see also* CALM (Commercial Advertisement Loudness Mitigation) Act 2012, US
low frequency emitter (LFE) 191
LtRt 192
LUFS (Loudness Units, Full Scale) 191–192

mag dubbers 100–101, *102*; Magnatec *102*, 138
Maguire, Ryan 95, 96, *110*

Mangini, Mark 89–90, 91, 100, 106, 114, 122, 123, 151, 152, 153, 163, 165, 178
mapping a plan 122–123
marketing 135
Martini 192
masking; frequency masking 22–23; temporal masking 23–24
master fader/master 192; *see also* submaster
mastering 161–163; mixing and mastering for the medium 163–166
McBryde, Mark 94, 97, *145*
MDE *see* DME (MDE, MED) (Dialogue, Music and Effects)
M&E (Music and Effects) track 97, 133, 176
mechanical sounds 98–99, 129
mechanics of sound; phase *31*, 31–33, *32*, *33*, *34*; reflections 34–37, *36*
MED *see* DME (MDE, MED) (Dialogue, Music and Effects)
Mickey-mousing 192
microphones 42–48, *43*, *44*, *45*, *46*, 73, 78–81, *79*, *80*, 84; ADR (Automated Dialogue Replacement) 140–141; amplitude 14; dialogue 19; dynamic microphone design 42–43, *43*; frequency response 18–19, *19*; objective hearing 41; phase interference 32–34, *33*, *34*; proximity effect 47, 48, 196; velocity 25; *see also*boom microphones; condenser (capacitor) microphones; lavaliere microphones; pickup (polar) patterns; shotgun microphones
Microsoft 201
"mini mixes" 165; *see also* submix, premixing
mix stage 155–160, *157*, *158*, *159*, 161–163, 192
mixdown 192, 198
mixer 192–193; *see also* production mixer
mixer interface/mixing control surface 54, *77*, 77–78, 193; *see also* DAWs (digital audio workstations)
mixing 103, 163–166, 193; evaluation and specifications 166–171, *169*; mixing and mastering for the medium 163–166; monitoring modes 171–174, *172*, *173*
modulation effects 62
monitors 193
mono (monaural) **64**, **65**, **66**, **67**, 105, 174, 192, 193
MOS ("Mit Out Sound") 88, 193
MP3 (MPEG-1 Audo Layer III/MPEG-2 Audio Layer III) 162, 193
MP4 (MPEG-4 Part 14) 193
MPEG (MotionPicture Editors Guild), IATSE Local 700 193–194, 202
MPEG (MotionPicture Experts Group) 193–194
MPSE (Motion Picture Sound Editors) 203
MS Pair (Mid-Side Pair) 194
multiband 194
multichannel delivery 176
multitrack 49, 51, 54, 75–76, 155, 173, 194
multitrack recorders *see* production recorders
Murch, Walter 5–6
music 116, 133–135, 136–137; mastering 161–163; music libraries 134, 194; music tracks 105, 113, 135; *see also* DME (MDE, MED) (Dialogue, Music and Effects); M&E (Music and Effects) track
Music Department 156, 194; coordination with 151–153; mix stage 155; music editors 135, 194; music supervisors 134, 150, 194

Nagra reel-to-reel field recorder 50, *50*
narrative (voiceover) sound 116
NBC 178
Netflix 7, 63, 175, 178
Neumann U87ai microphone 45, *45*
NIOSH (National Institute for Occupational Health and Safety), noise tolerances **15**, 15–16
Nixon, Richard 182
noise gates 60, 194–195
noise processing/reduction 138, 195; *see also* de-noise
noise processors 62
noise-cancelling headphones 31–32, *32*
Nolan, Christopher 93, 127, 160, *160*
non-diegetic sound 116, 187–188, 195; *see also* diegetic sound

Obama, B. 167
omnidirectional pickup (polar) pattern 46, *46*, 48, 185, 195
OSHA (Occupational Safety and Health Administration), noise tolerances **15**, 15–16
Ottoson, Paul N. J. 29–30

Pandora 163
"pasta treatment" 164
PBS 178
pebble ripple analogy 11, *11*, 12
perception *38*, 38–39
phase *31*, 31–33, *32*, *33*, *34*
Phonak Corporation 188
pickup (polar) patterns, microphones 45–48, *46*, 195; bidirectional 46, *46*, 47, 184, 185; omnidirectional 46, *46*, 48, 185, 195; unidirectional (cardioid) *46*, 46–47, 48, 185, 196, 201
pink noise *21*, 21–22
pitch 10; pitch-changing algorithms 62
playback 195
plug-ins *see* signal processors and plug-ins
podcasts 178–179
polar patterns *see* pickup (polar) patterns, microphones
polyWAV file format 195; *see also* WAV file format
post-production 88–103, *89*, *95*, *101*, *102*, 126; audio recording 50–51; dialogue editing 50–51, 92–93; Foley 94–97, *95*, 98; loudness measurement 170–1; mixing (re-recording, dubbing) 103; noise reduction 23, 24; predubbing 99–103, *101*, *102*; SFX editors 97–99; sound reflections 37; sound supervisor/sound designer 90–92, 105; sound team 88–90, *89*; transition from production 135–137
"p-pop" 47
predubbing 99–103, *101*, *102*, 113, 155, 165, 188, 195
pre-lay 196
premixing 156–157, 163, 165
preproduction 196
previz 196; *see also* storyboarding
Pro Sound Effects 147, 149
Pro Tools 104–105, *107*, *108*, 109, 196; dialogue edit *128*; Foley cue *148*; Pro Tools S6 console 55, *55*; SFX tracks *131*
production audio 68–71; best practices 85–88; equipment 74–85, *75*, *76*, *77*, *79*, *80*, *82*, *83*, 104–114, *107*, *108*, *110*, *111*, *112*, *113*; personnel 71–74; power management issues 81; transition to post-production 134–135; video monitoring capacity 81; *see also* post-production
production mixers 69, 70, 71–72, 74, 81–82; best practice 85, 86, 87; ownership of equipment 83–84
production recorders 48–54, *49*, *50*, *53*, *75*, 75–76, *76*, 84; digital recorders 48; flash media recorders 48, 49; sample rate and bit depth 51–54, *53*; tape-based recorders 48–49, *49*
props; Foley sound 96, *111*, 132; noise reduction 70–71, 143–144
Props Department 70, 144
ProRes 196
proximity effect 47, 48, 196
psychology of sound *see* sound, reality and psychology of

QuickTime file format 196; *see also* MP4 (MPEG-4 Part 14)

radio frequency interference (RFI) 78, 198, 200
radio, mixing and mastering for 162
rarefaction *10*, 11
redubbing 97, 132–133
reflections 34–37, *36*; and ADR (Automated Dialogue Replacement) 142–143
reverberation 36–37, 61, 62, 196
RFI (radio frequency interference) 78, 196, 200
ribbon element microphones 45
Royal Albert Hall, London 61, *61*

sample rate 51–52
SAP (Second Audio Program and Assistive Listening) tracks 174

scouts (location scouts) 197
screenwriting 123
selective hearing 40–41
Sennheiser MKE 600 microphone *44*, 45
Sennheiser MKH 416 microphone *79*
Set Decoration Department 70
SFX (sound effects) 113, 129–130, *131*, 132–133, 156, 197; ambient sounds 98, 99, 113; hard effects 98, 129, 149, 155, 191, 197; placement sound effects 136; predubbing 100; soft effects 98, 129, 149, 197; sound effects editors 90, 97–99, 124, 144–145, 149, 150, 152; sound effects libraries (stock effects) 94, 96, 149, 150, 197, 203; *see also* DME (MDE, MED) (Dialogue, Music and Effects); Foley; M&E (Music and Effects) track
shotgun microphones 47, 78, 199; *see also* boom microphone; unidirectional microphones
Showtime 178
Shure SM 58 microphone *43*
signal processors and plug-ins 56–62, *57*, *58*, *59*, *60*, *61*, 109
silence 14
"sits" 147
smartphones 63
social media 7, 178
soft effects 98, 129, 149, 197
software 106, 144
sound; envisioning of 153–153; nature of 2; reality and psychology of *38*, 38–42; in storytelling 2–5
sound carts 70, 76–77, *82*, 82–83, 198; *see also* follow carts
sound categories 117; commentative sound 120–121, *121*, 130, 186; contextual sound 117–118, *118*, 146, 186; descriptive sound 119–120, *120*, 125, 187
"sound collages" (Murch) 6
Sound Department/sound crew 68–71; best practices 85–88; communication between 73–74, 81–82, 85–86; multiple 88; *see also* boom operator; sound supervisors; third (cable person, sound utility)
sound design; applied design and workflows 115–160; overview of 5–7, *6*; scope and definition of 151
sound designers 91–92
Sound Devices 888 field recorder *76*
Sound Devices CL-16 control surface *77*
sound editing, working in 137–151, *141*, *142*, *145*, *148*, 149, *150*, 150–151
sound effects *see* SFX (sound effects)
Sound Ideas 149
sound mixers *see* production mixers
sound pressure level (SPL) 13–14, 197
sound supervisors 6, 90–92, 105, 122, 154; delegation 154, 155; and dialogue editing 127; mix stage 155; and music 151; SFX (sound effects) 144, 150; workflows 123–124
sound types 115–116
sound utility *see* third (cable person, sound utility)
sound waves 8–9; amplitude 12, *13*, 13–17, 184, 186; direct sound 35, 36, *36*; early reflected sound 35, 36, *36*; frequency *10*, 10–12, *11*; frequency response 17–24, *18*, *19*, *20*, *21*, 190; indirect/reflected sound 35–36, *36*; velocity 24–30, *26*, *27*, *29*; wavelength 30–31, 201
soundbars 63
SoundMiner 149
soundtrack channel permutations *64–67*
speakers, mix monitoring modes 171–174, *172*, *173*
spectrum processors 57–58, *58*
speech 39–41
speed of sound *see* velocity
SPL (sound pressure level) 13–14, 197
Spotify 163
spotting 94–95, 198; *see also* cue sheets
stems 126, 133, 198
stereo **64**, 105, 106, 158, 174, 177, 192, 198
stock music; *see also* music library
storyboarding 198; *see also* previz
storytelling, sound in 2–5
streaming platforms 7, 62, 63, 157, 175, 176–177, 178–179, 181, 198–199; mixing and mastering for 162, 163, 164
subgroups 199

submaster 199; *see also* master fader/ master
submix 199
subwoofers 74, 178, 199
surround/surround sound 174, 199
synchronous sound 117

technicolor 177
technology 180–181; *see also* equipment
television; terrestrial/broadcast 62, 63, 157, 161, 166–168, 175, 176–177; *see also* streaming platforms
temporal masking 199
temporal shift 15–16, 200
theater sound 5, 27, *27*, 37, 175
third (cable person, sound utility) 69, 70, 73, 197–198
thumpers 74, 199
THX 200
TikTok 7, 178
time processors 61–62, *62*
tinnitus 17
Todd-AO 141
Town, The 87
TPK (True Peak) 169, 200
tracks 200; track counts 155, 164
transducers 200
transient 200
tree 200

undipped deliverables 133, 177, 200–201; *see also* dipped
unidirectional (cardioid) pickup (polar) pattern *46*, 46–47, 48, 185, 196, 201
Universal Studios 94

vehicle sounds 98–99, 129
velocity 201; sound waves 24–30, *26*, *27*, *29*; and wavelength 30–31
VFX team 89, 90
vibration, frequency 10
video monitoring capacity, production audio 81
Villeneuve, Denis 89–90, 123–124, 152
vinyl 162, 163
VO Booth 109–112, *112*
voiceover (narrative) sound 116
VR (virtual reality) 179, 201

walkers 146, 189, 201
WAV (Waveform Audio File Format) 201; mixing and mastering for 162; *see also* BWAV file format
wavelength 30–31, 201
Waves 138
Weingarten, Mark 160
WGBH A Mix with S6 console *158*
white noise 19–21, *20*, 22
Williams, Tom 71, 72, 73–74, *82*, *83*, 84–85, 87
wireless 51, 73, 75, 78, 81, 84, 188, 198, 201
wires 201; phase interference 34
woofers **65**; *see also* LFE (low frequency emitter); subwoofers
workflows 123–126, *125*
workstations 104–106, *107*, *108*, 108–109, 202; portable 106, 108–109; *see also* DAWs (digital audio workstations)

XLR (External Line Return) 202

YouTube 7, 175, 178

Zaxcom DEVA 24 recorder 49, *50*, *75*
Zaxcom Mix-16 recorder 49–50, *50*, *77*
Zoom F8n Pro recorder 49, *49*, *76*

For Product Safety Concerns and Information please contact our EU
representative GPSR@taylorandfrancis.com
Taylor & Francis Verlag GmbH, Kaufingerstraße 24, 80331 München, Germany

www.ingramcontent.com/pod-product-compliance
Ingram Content Group UK Ltd.
Pitfield, Milton Keynes, MK11 3LW, UK
UKHW022017180926
475701UK00005B/895

* 9 7 8 1 0 3 2 4 4 6 2 0 2 *